Lazarus Fletcher

The Optical Indicatrix and the Transmission of Light in Crystals

Lazarus Fletcher

The Optical Indicatrix and the Transmission of Light in Crystals

ISBN/EAN: 9783337186173

Printed in Europe, USA, Canada, Australia, Japan

Cover: Foto ©berggeist007 / pixelio.de

More available books at **www.hansebooks.com**

THE OPTICAL INDICATRIX

AND

THE TRANSMISSION OF LIGHT IN CRYSTALS.

LONDON:
PRINTED BY WILLIAMS AND STRAHAN.
7 LAWRENCE LANE, CHEAPSIDE.

THE
OPTICAL INDICATRIX

AND THE

TRANSMISSION OF LIGHT IN CRYSTALS.

BY

L. FLETCHER, M.A., F.R.S.,

KEEPER OF MINERALS IN THE BRITISH MUSEUM;

FORMERLY FELLOW OF UNIVERSITY COLLEGE, MILLARD LECTURER AT TRINITY COLLEGE,
AND DEMONSTRATOR AT THE CLARENDON LABORATORY, OXFORD.

London:
HENRY FROWDE,
OXFORD UNIVERSITY PRESS WAREHOUSE,
AMEN CORNER.
—
1892.

CONTENTS.

	PAGE.
INTRODUCTION	ix

CHAPTER I. RECENT CHANGE OF VIEW AS TO THE PROPERTIES TO BE ASSIGNED TO AN ELASTIC LUMINIFEROUS ETHER.

Deduction of the form of the wave-surface for biaxal crystals	2
Its singularities	4
Dynamical difficulties of Fresnel's theory of double refraction	4
Results of the rigorous calculation of the vibratory motion of an elastic solid	5
Lord Kelvin's version of the elastic theory	6
Fresnel's line of reasoning, and the terms based upon it, must be abandoned	7
The form of the wave-surface for biaxal crystals was really discovered in another way	8

CHAPTER II. EVOLUTION OF THE OPTICAL INDICATRIX.

General nature of light	9
Light is due to change of state of matter	9
An ether is necessary	9
Permeation of ordinary matter by the ether	9
The change of state is periodic	10
The characters of an undulation	10
Light is an undulatory phenomenon	10
Sound is also an undulatory phenomenon	11
Intensity of light depends on the amplitude, colour on the period, of the vibration	11
Polarisation of light: plane of polarisation: transverse plane	11
Transmission of plane-polarised rays in glass and analogous media	12
Geometrical representation of the characters of a ray of plane-polarised light	12
The laws of ordinary reflection and refraction accounted for by an undulatory theory	13
The laws of refraction of light by a crystal of calcite accounted for by an undulatory theory	14
The wave-surface is identical with the ray-surface	14
Ray-front	15
In the case of calcite, the ray-surface has two sheets, and consists of a sphere and spheroid	15

	PAGE
Plane-polarisation of each of the refracted rays	16
The plane of polarisation is related to the radius vector of the ray-surface	17
The above might have led to the recognition of the possible existence and optical characters of biaxal crystals	17
Another mode of geometrically representing the characters of the extraordinarily refracted ray, by reference to the same spheroid, naturally presents itself	18
The same mode also suffices to represent the characters of the ordinarily refracted ray without necessitating the use of a second surface	19
The characters of the refracted rays can be simply expressed by reference to the spheroid alone	19
Generalisation	20
The Optical Indicatrix	20
Relation of the optical indicatrix to the general symmetry of the crystal	21

CHAPTER III. NATURALNESS OF THE METHOD.

Objections	22
The development of Fresnel's theory	23
Preliminary attempts at generalisation	24
The history of the ray-surface	26
The true nature of the luminiferous ether	28
The educational difficulty	29
Three other modes of generalisation	29
Advantages of the method here suggested	30

CHAPTER IV. DEDUCTION OF THE OPTICAL CHARACTERS CORRESPONDING TO AN ELLIPSOIDAL INDICATRIX.

ART.		
	General Relation	31
1-3.	The ray-surface: its construction, symmetry, and sections by the symmetral planes	32
4-6.	To find the characters of a ray in terms of the co-ordinates of R, the corresponding point on the indicatrix	36
7.	The equation of the ray-surface	37
8-9.	Given the direction-cosines of a line of transmission, to find (1) the velocities of the corresponding rays, (2) the co-ordinates of the corresponding points on the indicatrix	38
10.	The corresponding points R_1, R_2, are in a plane conjugate to the line of transmission	39
11.	To find their positions in the conjugate plane	39
12.	Perpendicularity of the planes of polarisation of the two rays transmissible along the same line	40
13.	Op, OR_1, OR_2, form a conjugate triad	41
14.	Given r_1 and r_2, to find $\lambda \mu \nu$	42
15.	Given the line of transmission, to find r_1 and r_2	43
16-20.	The optic bi-radials (secondary optic axes)	43
21-22.	The ray-front corresponding to the ray Or is perpendicular to the transverse plane $RNOr$, and intersects that plane in a line parallel to OR	48
23.	The velocity of the ray Or, resolved normally to the ray-front, is measured by the inverse of OR	52

ART.		PAGE
24-5.	OR is in general an axis of the section of the indicatrix by a plane parallel to the corresponding ray-front . . .	53
26.	The two rays corresponding to a given direction of front-normal	54
27.	The two front-normals corresponding to a given direction of ray	55
28-31.	Inclination of a ray to its front	55
32.	To find the direction-cosines of a front-normal in terms of the co-ordinates of R, the corresponding point on the indicatrix .	57
33.	Given $l\,m\,n$, the direction-cosines of the front-normal, to find f_1 and f_2	58
34.	Given f_1 and f_2, to find $l\,m\,n$	58
35-6.	Given $l\,m\,n$ or f_1 and f_2, to find the co-ordinates of the corresponding points on the indicatrix . . .	59
37.	Given $l\,m\,n$, to find the direction-cosines of the corresponding rays	60
38.	Given $\lambda\,\mu\,\nu$, to find the direction-cosines of the corresponding front-normals	60
39.	The front-normal surface	60
40.	The surface of wave-slowness or index-surface : it belongs to the same family as the ray-surface . . .	61
41-44.	The optic bi-normals (primary optic axes) . .	62
45-52.	The bi-radial and bi-normal cones . . .	66
53.	Representative surfaces derived from the indicatrix .	71

CHAPTER V. VARIOUS OPTICAL RELATIONS WHICH ARE INDEPENDENT OF THE PHYSICAL CHARACTER OF THE PERIODIC CHANGE.

1.	Object of the Chapter	73
2-8.	Preliminary representation of the periodic change at any point of a ray of light	73
9.	Discrepancy of observed and calculated results . .	83
10-12.	Its explanation	83
13-16.	A representative force : it is dependent on the luminous source .	85
17.	A fallacy	88
18.	In general, if a plane-polarised ray is transmissible in a given direction, the plane of polarisation can have at most two different directions	89
19.	The refraction cannot be higher than double . .	90
20.	Degree of the equation of the ray-surface . .	90
21.	The transmissibility of even a single plane-polarised ray is not a physical necessity : but if one position of a plane of polarisation is possible, there is a second at right angles with the first	90
22.	The transmission of a ray along an axis of tetragonal or hexagonal symmetry	91
23.	The velocity-factor	91
24.	It is necessarily the same for all directions perpendicular to an axis of tetragonal or hexagonal symmetry . .	91
25.	Transmission of a ray in a direction, lying in a plane of general symmetry, but oblique to an axis of tetragonal or hexagonal symmetry	92
26.	Transmission of rays along the axes of symmetry of an orthorhombic crystal	93

ART.		PAGE
27.	Transmission of rays in a symmetral plane of an ortho-rhombic crystal	93
28.	Intersections of the ray-surface with the symmetral planes of an ortho-rhombic crystal	94
29.	General equation of the ray-surface for an ortho-rhombic crystal	94
30.	The ray-surface for a mono-symmetric or anorthic crystal	95
31.	The form of the ray-surface is independent of the physical character of the periodic change	96
32.	Resilience	97
33-35.	Free and forced vibrations: simple cases	97
36.	Transmission of a simple forced vibration in an isotropically resilient medium	98
37-8.	More general cases of free or forced vibration of an æolotropically resilient medium	99
39.	Transmission of a simple forced vibration in an æolotropically resilient medium	101
40.	Case of an ortho-rhombic crystal	102
41.	Comparison with Fresnel's elastic forces	107
42.	Case of a mono-symmetric or anorthic crystal	107
43.	An unsatisfactory variation of Fresnel's method	107
44.	Transmission of elliptically or circularly polarised rays	109
SUMMARY		110

INTRODUCTION.

A SHORT account of the origin and development of this Tract, reprinted from the *Mineralogical Magazine*, may be of interest to the student.

It is known that many years ago Mr. Maskelyne undertook the writing of a Treatise on Crystallography. The book, embodying the mode of treatment of the subject adopted in his professorial lectures at Oxford and in a course of lectures given by him in the year 1874 to the Fellows of the Chemical Society, has been for some time complete as regards the purely crystallographic portion, and the methods and nomenclature employed in it are now familiar to the students of Crystallography at the University of Cambridge and at the City and Guilds of London Technical Institute, South Kensington; moreover, certain chapters on Crystallographic Physics have long been far advanced. But Mr. Maskelyne hesitated to publish his work without introducing, in the case of certain of the problems of Crystallographic Physics, some more simple and satisfactory treatment than any hitherto suggested; and, finding the subject too large to be satisfactorily dealt with in occasional hours of increasingly engrossing public occupations, he has from time to time invoked the aid of certain of his old pupils, among whom I have the happiness to be numbered.

Some years ago I responded to a call of this kind, which involved the investigation of the behaviour of a crystal, viewed merely as an æolotropic body, in its relations to change of temperature. At that time the permanent rectangularity of a definite triad of lines of a mono-symmetric or anorthic crystal was very generally accepted by crystallographers. In two papers, read before the Crystallological Society (1879-83), this view was criticised, and the whole subject of the dilatation of crystals on change of temperature was discussed by the aid of mathematics and general reasoning of an extremely simple character, as compared with any which had been recorded by previous workers.[1]

The present Tract deals with another and more important problem, that, namely, of the behaviour of a crystal in respect to the refraction of light. The beautiful process invented by Fresnel has long been recognised, and increasingly so as time has gone on, as being dynamically unsound; but there was no more rigorous method which did not involve mathematics of too high an order to be introduced into a book on Crystallography that should be generally useful, and none at all which was completely concordant with experiment in its results. The task proposed to me by my old friend and teacher was that of presenting the subject of refraction in a far simpler form to the student; and, as an almost necessary condition for such a presentment, I had to endeavour to approach the subject by such a path as would render it possible to avoid, as much as possible, any discussion of the characteristics of the ethereal medium; for physicists were still uncertain, not merely as regards the properties to be assigned to an elastic luminiferous ether, but even as regards the physical character of the pulsation which constitutes light.

Meantime the problem assumed a different aspect; for at the end of 1888 Lord Kelvin remarked that incompressibility, hitherto regarded as absolutely indispensable, is really unnecessary to the stability of the ether; and he showed that the laws which determine the intensities of ordinarily reflected or refracted light are deducible from the properties of an ether which is compressible for the forces concerned in the transmission of light; Mr. Glazebrook immediately followed with proofs that other important phenomena (such as ordinary and anomalous dispersion, double refraction, and metallic reflection) are likewise consistent with the new version of the elastic theory.

Fresnel's process thus became more untenable than ever, even for the mere correlation of optical facts, for its hypotheses are completely at variance with those required by the new version. An elastic luminiferous ether is now to be regarded as compressible instead of incompressible; its effective elasticity (of figure) is to be regarded as constant instead of variable, its effective density as variable instead of constant, for different media, and for different directions in the same medium if the latter be bi-refractive.

But the mathematical development of the new version, involving as it does the idea of a slipless rigid boundary and a variable effective ethereal density, and the use of partial differential equations and triple integrals, calls for an amount of special knowledge which it is impossible to require from a purely crystallological student. It thus remained to invent, if possible, a process which should involve only elementary mathematics, be consistent with the

---

[1] *Lond. Dub. and Edinb. Philos. Magazine;* 1880, ser. 5, vol. 9, p. 80: 1883, ser. 5, vol. 16, pp. 275, 344, 412.

hypotheses and results of the new version of the elastic theory, and yet be sufficient to serve the purpose of correlation of the phenomena of refraction, for which the method of Fresnel has been found of so great service.

In the search for such an elementary process my attention was attracted to several remarkable facts:—

1st. The direction *now* assigned to the ethereal vibration of a given ray is identical with that of Fresnel's elastic force.

2nd. The velocity of transmission of the *ray* is proportional to the magnitude of that force.

3rd. The force is represented in *direction* by the normal of the "ellipsoid of elasticity," and in *magnitude* by the inverse of the length of the normal intercepted by the ray.

It seemed that so simple a set of relations must be capable of translation by the aid of elementary mathematics into mechanical ideas consistent with the new hypotheses.

An attempt to effect this took the form given on page 107, but was discarded as unsatisfactory. I became gradually convinced that all such attempts are premature, and that the adoption of any method of the kind would involve a possible recurrence of the present difficulty: for it is far from established that the new version, though incomparably more concordant with experimental results than the old, is anything more than a mere mechanical analogy, liable at any moment to be found incomplete, and to be replaced by a version of a totally different character. In fact, recent experiments seem to have established that light-waves and electro-magnetic waves only differ in length, yet the latter are deemed inexplicable as mere vibrations of an elastic ether.

By this time, however, it had become manifest that a simple method of generalisation would have directly led to the discovery of the optical characters of biaxal crystals, without any reference to the specific characters of the ether at all; but the difficulty remained that the general form of the wave-surface had apparently been arrived at, not in this simple way, but by *a priori* reasoning founded on the properties of an elastic incompressible ether. This belief is a very general one, and is an almost inevitable result of a study of Fresnel's memoir, as published in the Transactions of the French Academy. That the belief is a mistaken one will be evident from the detailed history given in Chapter III. It must be remembered, in reading the memoir, that Fresnel, who died before its actual issue, was contending for the undulatory as against the emissive theory of light, and that most of his remarks are applicable to vibrations in general as well as to the motions of the parts of an elastic ether.

When it is made clear that Fresnel's deductive process was really an *a posteriori* one, and had not led to the discovery of the general form of the wave-surface, it is possible to part from his theory of double refraction with less reluctance; and in adopting the method here suggested we shall merely be reverting to one which is at least analogous that by which his discovery was actually made. And this can be done notwithstanding the admiration for Fresnel's brilliant researches which must be felt by every reader of his various memoirs. When those researches began, in 1815, the emissive theory of light was in the zenith of scientific favour; when Fresnel died, in 1827, active opposition to the undulatory theory had virtually ceased: a result which was a direct consequence of Fresnel's reasoning and discoveries.

As regards the method here suggested, it will be found that the idea of a correspondence between the characters of a ray and the geometrical characters at a point on an ellipsoid brings great simplicity into the study of the optical characters of crystals; a simpler relationship than that which is taken as the basis of Chapter IV could not be desired.

As for the mathematical development of that Chapter, it need only be pointed out that the equation of the ray-surface is deduced without the aid of the differential calculus or any complicated method of elimination: indeed, the Chapter requires no higher mathematical knowledge than is implied in the idea of conjugate diameters of an ellipsoid; the knowledge of infinitesimals demanded for the geometrical solution given in Art. 21 of that Chapter being of a very elementary character. The investigation of the geometrical relations of Fresnel's wave-surface was long ago exhausted by Hamilton, Mac Cullagh, Sylvester, Plücker, and Cauchy; yet it is hoped that the adoption of the basis here suggested, the use of rays instead of waves, and the comparatively elementary character of the mathematical treatment given in Chapter IV, may enable the crystallological student to acquire a clearer idea of the geometrical relations of the wave-surface, and of the optical characters of crystals in general, than he is able to attain to by means of a method which is based on the hypothesis of varying ethereal density.

It may assist the memory of the student if it is remarked that many of the symbolic letters used in the notation of Chapter IV are the initials of the corresponding words: r R_1 R_2 ρ are all related to rays, N to a normal, f to a front, p_1 p_2 $[p_1]$ $[p_2]$ π_1 π_2 and s_1 s_2 $[s_1]$ $[s_2]$ σ_1 σ_2 to the so-called primary and secondary optic axes respectively.

In the last and more difficult Chapter the object in view is a different one: it is there sought to deduce the general form of the wave-surface by elementary reasoning from simple hypotheses relative to the general characters of undulations, and without assuming that the vibration is an actual motion due to ethereal elasticity.

L. FLETCHER.

London, March 31st, 1891.

ON THEORIES OF LIGHT.

"Most thinkers of any degree of sobriety allow that an hypothesis of this kind is not to be received as probably true, because it accounts for all the known phenomena. But it seems to be thought that an hypothesis of the sort in question is entitled to a more favourable reception, if, besides accounting for all the facts previously known, it has led to the anticipation and prediction of others which experience afterwards verified. Such predictions and their fulfilment are, indeed, well calculated to strike the ignorant vulgar. But it is strange that any considerable stress should be laid upon such a coincidence by scientific thinkers. If the laws of the propagation of light accord with those of the vibrations of an elastic fluid in as many respects as is necessary to make the hypothesis a plausible explanation of all or most of the phenomena known at the time, it is nothing strange that they should accord with each other in one respect more. Though twenty such coincidences should occur, it would not follow that the phenomena of light are results of the laws of elastic fluids, but at most that they are governed by laws in some measure analogous to these; which, we may observe, is already certain, from the fact that the hypothesis in question could be for a moment tenable. Who knows but that some third hypothesis, including all these phenomena, may in time leave the undulatory theory as far behind as that has left the theory of Newton and his successors?" (*Mr. J. S. Mill*, 1843-51.)

"We are led to the conception of a complicated mechanism capable of a vast variety of motion, but at the same time so connected that the motion of one part depends, according to definite relations, on the motion of other parts, these motions being communicated by forces arising from the relative displacement of the connected parts in virtue of their elasticity. The agreement of the results seems to show that light and magnetism are affections of the same substance, and that light is an electro-magnetic disturbance propagated through the field according to electro-magnetic laws." (*Prof. J. Clerk Maxwell*, 1865.)

"I only mean that *if* light, as is generally supposed, consists of transversal vibrations similar to those which take place in an elastic solid, the vibration must be normal to the plane of polarisation. There is unquestionably a formal analogy between the two sets of phenomena extending over a very wide range; but it is another thing to assert that the vibrations are really and truly to-and-fro motions of a medium having mechanical properties (with reference to small vibrations) like those of ordinary matter." (*Lord Rayleigh*, 1871.)

"While the elastic-solid theory, taken strictly, fails to represent all the facts of experiment, we have learned an immense amount by its development, and have been taught where to look for modifications and improvements. Nor is it surprising that a simple-elastic-solid theory should fail. The properties we have been considering depend on the presence of matter, and we have to deal with two systems of mutually interpenetrating particles. It is clearly a very rough approximation to suppose that the effect of the matter is merely to alter the rigidity or density of the ether. The motion of the ether will be disturbed by the presence of the matter; motion may even be set up in the matter-particles. The forces to which this gives rise may, so far as they affect the ether, enter its equations in such a way as to be equivalent to a change in its density or rigidity, but they may, and probably will, in some cases do more than this." (*Mr. R. T. Glazebrook*, 1885.)

"It follows that the luminiferous ether is experimentally shown to be the medium to which electric and magnetic actions are due, and that the electro-magnetic waves are really only very long light-waves. If magnetic forces are analogous to the rotation of the elements of a wave, then an ordinary solid cannot be analogous to the ether, because the latter may have a constant magnetic force existing in it for any length of time, while an elastic solid cannot have a continuous rotation of its elements in one direction existing within it. The most satisfactory model, with properties quite analogous to those of the ether, is one consisting of wheels geared with elastic bands." (*Prof. G. F. Fitzgerald*, 1890.)

CHAPTER I.

Recent Change of View as to the Properties to be assigned to an Elastic Luminiferous Ether.

Deduction of the form of the wave-surface for biaxal crystals.

FRESNEL'S representation of the laws of transmission of rays of light in biaxal crystals, by reference to the surface distinguished by his name, has long been regarded as one of the greatest achievements in the domain of Physical Science. In his memoir[1] on Double Refraction, Fresnel proceeded as follows:—

1. He assumed that the transmission of a ray of light is effected by means of an elastic ether vibrating transversely to the ray-direction.

To the ether is thus assigned a property not belonging to a perfectly fluid body in a state of rest: perfect fluidity of a body at rest involves incapacity of resistance to mere change of shape, and it is to such distortional resistance that transverse vibrations must be due.

2. He assumed that the ether of a crystal, when undisturbed, is a system of equal particles, in stable equilibrium under their mutual attractions; and that, for each pair of particles, the attraction depends solely on some function of the distance between them and acts in the line joining the centres.

He showed that in a medium so constituted there are at least three directions, at right angles to each other, such that the force necessary to the maintenance of a small displacement of a single particle of the ether along any one of them will act in the line of the displacement, and be proportional to it in magnitude: that the elastic force evoked by the displacement of a single particle of the ether through unit-distance along each of these directions may be different, say a^2, b^2, c^2, respectively: that in this case, which is assumed to be that of the

---

[1] *Mémoires de l'Acad. de l'Institut de France*, 1827, vol. 7, pp. 45-176.

ether in a biaxal crystal, the elastic force due to the displacement of a single particle of ether in any direction distinct from the three already mentioned will act in a direction different from that of the displacement; that if the direction of a radius vector of the surface $a^2x^2 + b^2y^2 + c^2z^2 = (x^2 + y^2 + z^2)^2$ represent that of the displacement of an ethereal particle, and the corresponding elastic force for a displacement through unit-distance be resolved along and perpendicular to the line of displacement, the former component is proportional to the square of the radius vector in magnitude: that for displacements of a single particle in directions lying in a given plane passing through the centre of the above surface, the elastic force is generally obliquely inclined to the plane, but that there are always two directions, namely, those of the longest and shortest diameters of the section of the above surface by the given plane, for which the resolved component of the elastic force in the given plane acts in the line of displacement.

3. He assumed that the ether is virtually incompressible for the forces concerned in the transmission of light.

Neglecting, therefore, the component of the elastic force normal to a plane containing a set of similarly displaced particles (wave-front) as being without effect by reason of the incompressibility of the ether, Fresnel inferred that, for particles in the given plane, vibrations parallel to either the longest or shortest diameter of the corresponding section of the above surface must be persistent, since the only effective component of the elastic force for each particle then acts in the direction of the displacement.

4. From a suggested but forced analogy of a line of vibrating ether-particles to a vibrating string, Fresnel assumed that the velocity of transference of a wave-front along its normal is directly proportional to the length of that principal diameter of the section of the above surface by the wave-front which is parallel to the direction of vibration.

Hence finding, by the usual mathematical process, the envelope of planes representing the positions to which wave-fronts, with every possible direction, would arrive after the lapse of the same interval of time, Fresnel concluded that the wave-surface for a biaxal crystal is represented by the equation

$$\frac{a^2x^2}{r^2-a^2} + \frac{b^2y^2}{r^2-b^2} + \frac{c^2z^2}{r^2-c^2} = 0:$$

further, as the front corresponding to any ray is parallel to the tangent plane to the wave-surface at the point where the ray meets it, the vibration is, in general, obliquely, not perpendicularly, transverse to the direction of the ray.

5. Hence Fresnel also inferred that the velocities of the two rays which can be transmitted along a given direction are directly proportional to the axes of the ellipse in which the ellipsoid $\frac{x^2}{a^2}+\frac{y^2}{b^2}+\frac{z^2}{c^2} = 1$ is intersected by a plane normal to the common direction of the rays.

Singularities of the form.

The closed surface represented by the above equation is of very peculiar form, and consists of two concentric ellipsoid-like sheets, which are symmetrical with respect to three rectangular planes. There are four points common to both sheets; they are situated at the extremities of two diameters lying in one of the planes of symmetry: in the neighbourhood of each of these points the sheets are drawn towards each other, and the surface has there the shape of a double cone; an infinite number of tangent planes to the surface can thus be drawn at each of them. Further, two planes and their parallels respectively touch the surface, not at one point nor at two points, but at an infinite number of points which lie on the circumference of a circle.

These geometrical singularities of the wave-surface, first noticed by Sir William Hamilton five years after the death of Fresnel, point to the existence in biaxal crystals of certain optical characters which had up to that time remained undiscovered, and seemed too strange to be real: the establishment of their actuality by Lloyd has been regarded as the crowning triumph of Fresnel's theory of double refraction; for not only are the phenomena strange, but their observation demands a combination of circumstances which places them beyond the range of accidental discovery.

Dynamical difficulties of Fresnel's theory of double refraction.

As continued experiment and precise observation have served only to establish the high degree of accuracy of the form assigned to the wave-surface by Fresnel,[1] it might naturally be inferred that the assumptions which lead, after so elaborate a course of reasoning, to a surface presenting these singularities must be themselves beyond cavil. Yet, strange to say, the mathematical process, by which the surface is thus arrived at, is one of which the weakness was recognised by the author himself, and the

---

[1] Kohlrausch: *Wied. Ann.*; 1879, vol. 6, p. 86; vol. 7, p. 427.
Glazebrook: *Phil. Trans.*; 1879, vol. 170, part 1, p. 287. *Proc. Roy. Soc.*; 1883, vol. 34, p. 393.

theory has long been regarded as dynamically unsound; further, the characters assumed for the ether, though they lead to the true wave-surface, have since been found to have for necessary consequences other optical laws which are inconsistent with the results of experiment. On the other hand, the same form of wave-surface can be arrived at from other sets of assumptions, which have thus the same claim to recognition;[1] yet they are inconsistent with those of Fresnel, and with each other. As the later hypotheses which lead to Fresnel's wave-surface have been found to have other consequences which are contradicted by experimental results, the comparative simplicity and the historical interest of the method of Fresnel have sufficed to secure the adoption of his assumptions and corresponding terminology in the general literature relating to the optical characters of crystals.

The fact that Fresnel's wave-surface has been deduced from several inconsistent sets of assumptions as to the characters of the ethereal motion suggests that the form may really depend on the feature common to all, namely, the transmission of a periodic change of state differently related to different sides of the ray, and be otherwise independent of the physical character of the transmitted change: the suggestion is discussed in Chapter V.

Results of the rigorous calculation of the vibratory motion of an elastic solid.

The rigorous calculation[2] of the vibratory motion of the parts of an isotropic elastic solid is found to involve two quantities, which are generally denoted by A and B: the latter, B, measures the rigidity, or the resistance of the body to simple change of shape, or the *elasticity of figure;* the former, A, is connected with B, and with k (which measures the resistance to simple change of volume, or the *elasticity of volume*), by the relation $k = A - \tfrac{4}{3}B$. Further, it can be shown that a vibratory motion of the parts of an elastic medium generally gives rise to two kinds of waves, due respectively to distortional and condensational-rarefactional vibrations; the former travelling with velocity $\sqrt{\dfrac{B}{\rho}}$, the latter (which correspond to those of sound) with velocity $\sqrt{\dfrac{A}{\rho}}$, where ρ is the density of the medium. Now, if the transmission of light through a singly refractive medium be

---

[1] *e.g.* Challis in the *Trans. Camb. Phil. Soc.;* 1847, vol. 8, p. 524.
[2] A most valuable Report by Mr. Glazebrook on Optical Theories is published in the *Rep. Brit. Assoc.* for 1885, pp. 157-261.

due to the vibratory motion of an isotropic elastic solid, all the energy persists in the form of distortional vibrations perpendicular to the ray; hence the characters of the ether must be so assumed as to secure the absence of the condensational-rarefactional vibrations. For this purpose we may make either of two assumptions, namely, that A is virtually zero or that A is virtually infinite as compared with B: in the former case the condensational-rarefactional wave is got rid of by making its velocity zero; in the latter case by making the velocity infinite. But it was long believed that the former assumption was otherwise inadmissible: for it was supposed by Green and later mathematicians that the quantity $A - \tfrac{4}{3}B$ is necessarily positive, if the equilibrium of the parts of an elastic body is stable; and this is impossible if A is zero, for B is essentially a positive quantity: hence it only remained to assume A and therefore also k infinite, and thus the ether to be virtually incompressible.

Double refraction could then be consistently explained by a variation of the rigidity of the ether of a bi-refractive crystal with the direction; but it was necessary, for dynamical reasons, to assume the vibrations to be in, not perpendicular to, the plane of polarisation.

On the other hand, Lord Rayleigh[1] has proved that the phenomena due to the scattering of light by small particles require the vibrations of the ether to be perpendicular to the plane of polarisation; he has further shown that no theory based on varying rigidity can possibly be satisfactory, and that the variation of density in different directions in a biaxal crystal would lead dynamically to a form of wave-surface different from that of Fresnel, if the ether be incompressible for the forces involved in the propagation of the vibrations.

Lord Kelvin's version of the elastic theory.

From this position of dead-lock, according to which the ether must be both compressible and incompressible, the theory that the transmission of light is effected by the vibrations of an elastic medium has only recently been extricated. At the end of 1888 Lord Kelvin,[2] re-examining the problem of the stability of the equilibrium, found that the condition that $A - \tfrac{4}{3}B$ is a positive quantity becomes unnecessary, "provided we either suppose the medium to extend all through boundless space, or give it a fixed containing vessel as its boundary:" with either of these provisions, the stability only requires that A should not be negative, and it is therefore possible to get rid of the condensational-rarefactional wave by

---

[1] *Lond. Edin. and Dub. Philos. Magaz.*, 1871, ser. 4, vol. 41, p. 451.
[2] *Ibid.*, 1888, ser. 5, vol. 26, p. 414.

the assumption hitherto deemed inadmissible, namely, that A is zero; this involves the compressibility of the ether for the forces concerned in the propagation of light. As a mechanical illustration, Lord Kelvin points out that " homogeneous air-less foam held from collapse by adhesion to a containing vessel, which may be infinitely distant all round, exactly fulfils the condition of zero-velocity for the condensational-rarefactional wave; while it has a definite rigidity and elasticity of form, and a definite velocity of distortional wave, which can easily be calculated with a fair approximation to absolute accuracy."

Starting with the new assumption, Lord Kelvin was able to deduce correct expressions for the intensities of ordinarily reflected or refracted light: and Mr. Glazebrook[1] has since shown that the elastic theory in its new form fully accounts for dispersion, including anomalous dispersion (like that of cyanin), double refraction, and metallic reflection, and further that it leads to a correct expression for the velocity of light in a moving medium. According to the new version, the vibrations of the ether are perpendicular to the plane of polarisation, even in biaxal crystals, and thus always perpendicularly transverse to the ray: further, the matter-particles and ether-particles are supposed to react on each other: and if their vibrations are synchronous, the former may even be set in appreciable motion by the latter. As the reaction of the matter and ether may produce the same effect on the motion of the ether-particles as would result from a simple variation of the rigidity or density of the ethereal medium, it becomes convenient to distinguish between the *actual* and *effective* values of the rigidity and density.

It is clear that the new version of the properties of the elastic ether, whether really true or not,[2] is far more satisfactory than any hitherto suggested, and must replace the older versions until a better one is proposed. Hence it becomes necessary, for those who adopt an elastic ether as the basis of the undulatory theory, to regard (1) the ether as compressible, even for the forces concerned in the propagation of light; (2) the actual density and rigidity of the ether as identical for all bodies; (3) the effective rigidity as invariable; (4) the effective density as different in different bodies, and, in the case of doubly refractive crystals, in different directions within the same body.

Fresnel's line of reasoning, and the terms based upon it, must be abandoned.
For the great majority of mineralogical students, the chief value of the

---
[1] *Ibid.*, 1888, ser. 5, vol. 26, p. 521.
[2] *Ibid.*, p. 533; 1889, vol. 27, pp. 246, 253: *Nature*, 1889, vol. 40, p. 32.

hypothesis of an elastic ether is in the correlation of the phenomena observed when light is transmitted through crystals; for which purpose it is very desirable that the student should be able to reach the wave-surface, if practicable, by means of elementary reasoning based on observed facts of a simple character. The rigorous calculation of the motions of a vibrating elastic medium is not a simple process: it involves, indeed, mathematics of so high an order that the derivation of the wave-surface in this way will always be unintelligible to the ordinary student of crystals. On the other hand, the only comparatively simple mode of derivation of the wave-surface, as yet invented, that of Fresnel, depends upon assumptions of incompressibility and varying elasticity which are now deemed untrue; and further, involves for biaxal crystals a general obliquity of transverse vibration, not in accordance with the latest version of the elastic theory. Under present circumstances, the process of Fresnel, even if adopted on account of its great historical interest, must be puzzling to the student, and inevitably lead to the acquisition of wrong views as to the properties to be assigned to the luminiferous ether; hence it becomes necessary to abandon the whole process, and all those terms now in common use (ellipsoid of optic elasticity, axes of optic elasticity, coefficients of optic elasticity) which are based upon it.

The form of the wave-surface for biaxal crystals was really discovered in another way.

The great difficulty in the correlation of the phenomena of the transmission of light through biaxal crystals, as already stated, lies in the derivation of the wave-surface. The form of the surface is too extraordinary to be directly assumed either as a probable one *a priori*, or as suggested by experimental results. If it can be shown that the form of the wave-surface for biaxal crystals is suggested by a simple generalisation, independently of any particular version of the undulatory theory, and might have been brought in this way within the province of experimental investigation, the greater part of the present educational difficulty will be removed from the path of the student. In fact, we shall find that it was really by a process of generalisation, though not indicated in the composite memoir of 1827, that Fresnel himself was first led to the true form of the wave-surface for biaxal crystals. The properties of an incompressible elastic ether were mathematically developed by him after the discovery of the true form of the wave-surface had been made.

CHAPTER II.

Evolution of the Optical Indicatrix.

In the present Chapter it is sought to show that a certain surface, here termed the Optical Indicatrix, naturally suggests itself as a means of correlation of the laws of transmission of light in uniaxal crystals; a simple generalisation then suggests the possible existence of biaxal crystals, and the general nature of their optical properties. The reasoning may be arranged as follows:—

General nature of light.

Light travels with finite velocity.

A flash of light transmitted from one body to another may thus for a time be wholly in the intervening space; hence the transmission of light must be one either of matter or of change of state of matter.

Light is due to the change of state of matter.

Two rays of light of the same colour, travelling in the same direction along the same line, may annihilate each other.

Hence the transmission of light cannot be one of matter; it must be a transmission of change of state of matter, and the change must be capable of representation by positive and negative quantities.

An ether is necessary.

Light travels across interplanetary space.

Hence interplanetary space must be filled with one or more kinds of matter, capable of transmitting a particular kind of change of state with an enormous but finite velocity (186,000 miles a second), and for distances amounting to millions of millions of miles. We may conveniently assume that the extraordinary matter is wholly of one kind, and designate it by a special name, *ether*; it must be extremely subtle, for it offers no appreciable resistance to the motion of the planets.

Permeation of ordinary matter by the ether.

Light is transmitted, but with different velocities, through ordinary matter.

Hence either ordinary matter is itself capable of transmitting this particular kind of change of state, or it is permeated by an ether capable of so doing. Having regard to the enormous velocity with which light is propagated through interplanetary ether and different kinds of ordinary matter, we may assume that the same kind of ether is concerned in the transmission, and that the variation of velocity and other characters is due to the influence of the ordinary matter on the properties of the permeating ether.

The change of state is periodic.

If two rays, continually transmitted along the same line, annihilate each other, annihilation again takes place if either ray is transferred through any multiple of a certain measurable distance along the direction of transmission.

Hence, so long as a single ray of light is being transmitted along a line, the state of the ether at a given instant is the same at all points distant from each other by a certain measurable quantity, which we may denote by λ. But the continual uniform transmission of the change of state along the line involves a continual and periodic change of state at each point of the line; the duration of the period being the same at all points, and always equal to the time necessary for the transmission of the change through the distance λ along the line: if v be the distance of transmission during the unit of time, the period will thus be $\frac{\lambda}{v}$. During a single period, the ether at any point in the line of transmission experiences all those changes which belong at a given instant to all points in a length λ of the line of transmission.

The characters of an undulation.

Whatever be its physical nature, a periodic change of character at any point is termed a *vibration* of the character: its maximum value, the *amplitude* of the vibration: the interval of time required for a complete vibration, its *period*: the state at a given instant, the *phase* of the vibration: the relation between the phase and the time, the *law* of the vibration. If, further, the change is being transmitted along a line or *ray*, the configuration of the states at all points of the ray at a given instant is termed an *undulation*: the least part of an undulation which includes all varieties of phase is termed a *wave*, and the distance occupied by a wave, a *wavelength*.

Light is an undulatory phenomenon.

It follows from the above that, in this general sense, light is undoubtedly an undulatory phenomenon of some kind or other.

Sound is also an undulatory phenomenon.

By similar reasoning, it follows that sound is an undulatory phenomenon. Experiment shows that the transmission of sound is effected by ordinary matter, and that the change of character is one of oscillation of the material particles, the oscillation being generally solely in the direction of the transmission. The properties at any point of a line of transmission of a continued uniform sound, namely intensity, note and *timbre*, must depend on the characters of the vibration at the point, and thus on the amplitude, period and law: experiment proves that the intensity of a simple sound depends solely on the amplitude, and the note solely on the period.

Intensity of light depends on the amplitude, colour on the period, of the vibration.

Similarly, the corresponding properties at any point of a ray of ordinary light, intensity and colour, may be assumed to depend on the characters of the vibration at the point, and thus on the amplitude, period and law: we may tentatively assume, from analogy with sound, that the intensity of a simple ray depends solely on the amplitude, the colour solely on the period.

Polarisation of light; plane of polarisation; transverse plane.

But common light is capable of a change to which there is no parallel in the case of sound. A ray of common light transmitted through air acquires distinctive characters by reflection at a certain angle of incidence from a sheet of glass: as tested by reflection at the same angle of incidence from a second plate of glass, it has different properties on different sides; its properties being symmetrical, however, at every point of the path to the same two perpendicular planes intersecting in the ray: one of the planes of symmetry is the plane of incidence and reflection from the first plate. As the planes of symmetry of the ray are dissimilar and can be experimentally distinguished from each other, that which coincides with the plane of incidence and reflection may conveniently be termed the *plane of polarisation;* the second plane of symmetry may be distinguished as the *transverse plane.* A ray having the same characters, however induced, is said to be plane-polarised.

Hence the periodic change of the ether at any point of an aerially transmitted plane-polarised ray of light is not solely related to the direction of transmission, and thus differs in kind from that which characterises sound. For the suggestion of the laws of double refraction, preciser knowledge of the character of the change is unnecessary.

Transmission of plane-polarised rays in glass and analogous media.

If a plate of ordinary glass or any analogous medium be placed with its faces perpendicular to an aerially transmitted plane-polarised ray, the light which emerges from the glass is found to be still plane-polarised, and the position of the plane of polarisation is found to be unaltered whatever the thickness of the plate: this is still true, if the plate be turned through any angle round its own normal.

As the direction of the ray within the plate is coincident with the direction of the ray before incidence and after emergence, we may thus reasonably assume that, at all points of the line of transmission *within the plate itself*, the periodic change of the ether is symmetrical to the same two planes; in which case the position of the symmetral planes of the periodic change is wholly independent of the glass and depends only on the direction of the plane of polarisation of the incident ray. A plane-polarised ray transmissible in any direction within such a medium may have any azimuth of plane of polarisation whatever.

Geometrical representation of the characters of a ray of plane-polarised light.

In representing the transmission of a ray of plane-polarised light, of a

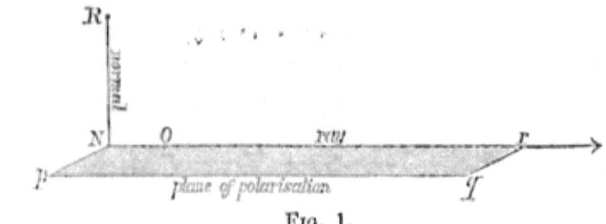

Fig. 1.

single given colour and given intensity, within a given medium, we have thus three characters to consider:—

    1. The line of transmission of the ray,
    2. The direction of the plane of polarisation,
    3. The velocity of transmission.

The direction of a plane being most conveniently defined by the direction of its normal, the above three characters may be geometrically represented by means of two intersecting perpendicular lines, one of them definite in position, the other only in direction: and any definite function of the length of either may represent the velocity.

The direction of transmission $O\,r$, and the plane of polarisation $O\,p\,q\,r$

of a given ray (Fig. 1), may thus be represented by two lines Or, RN; where RN is any line perpendicular to the plane $Opqr$, and therefore also to the ray Or: the velocity of transmission may be represented by any function of either of the lines Or, RN.

If O, a point on the ray, be given, and the normal of the plane of polarisation be taken to intersect the ray, all the characters may be represented by means of a *single line RN, not passing through the given point*: for the line Or is then known, since it passes through O and intersects RN perpendicularly.

The laws of ordinary reflection and refraction accounted for by an undulatory theory.

Two hundred years ago (1678-90), Huygens showed, by reasoning which is really independent of the physical nature of the periodic change, though he imagined it to be identical in character with that involved in the transmission of sound, that the laws of ordinary reflection and refraction of light are compatible with an undulatory theory. He assumed that a general disturbance of the ether at any given point must eventually produce disturbances at all other points of the medium, and that in a transparent body showing ordinary refraction the velocity of transmission of the disturbance is independent of the direction; all points on a spherical surface having the given point for centre are thus at any moment in a similar state of disturbance. If we have regard to the arrival of the disturbance from its origin, we may say that in this case the *front* of the disturbance at any epoch is a sphere. The front of the disturbance due to a single centre may, for the sake of brevity and generality, be called the *wave-surface*. If the disturbance at the centre be persistent and periodic, the surface which defines the front of the disturbance at a given epoch passes through points of the medium at which, notwithstanding the continual change at each point, there is persistent identity of phase of vibration.

Huygens gave a geometrical construction for the determination of the direction of the refracted ray by means of the spherical wave-surface, the direction being that of a line joining the point of incidence of the ray to the point of contact of a tangent plane of the wave-surface, drawn through an auxiliary line which lies in the refracting surface and is normal to the plane of incidence: if v be the velocity in the first medium, i the angle of incidence of the ray, and the size of the sphere correspond to the lapse of a unit of time, the distance of the auxiliary line from the point of incidence is $\dfrac{v}{\sin i}$.

The laws of refraction of light by a crystal of calcite accounted for by an undulatory theory.

In the case of calcite, the refraction is in general not single but double. One of the rays, and only one, follows the laws of ordinary refraction for all directions: hence Huygens inferred that the surface of disturbance corresponding to this ray is the same as for ordinary media, namely, a sphere. It was necessary to assume a different form of surface of disturbance to account for the extraordinary refraction of the other ray, and the surface which first suggests itself, after a sphere, is an ellipsoid: further, since the refraction of the second ray is the same for all directions equally inclined to a special direction in the calcite-crystal, or since rays lying in a plane perpendicular to this line obey the laws of ordinary refraction, it is necessary for the surface of disturbance to be one of revolution about that direction as axis. Testing this hypothesis and finding it satisfactory, Huygens inferred from his observations that the surface of disturbance corresponding to the second ray is really a spheroid, touching at the extremities of its axis the spherical surface of disturbance corresponding to the first ray. This relation between the surfaces has been confirmed by later experimenters and found to hold for other crystals analogous to those of calcite: it is undoubtedly a Law of Nature.[1]

The direction of the extraordinarily refracted ray is given by the same geometrical construction as before, the surface of disturbance being taken as a spheroid instead of a sphere.

The wave-surface is identical with the ray-surface.

Let rs, RS, (Fig. 2) be two wave-surfaces due to an origin O, and with the line OrR as axis describe a cone of small angle, determining areas ab, AB, upon them. There is great difficulty in imagining the exact nature of the physical process by which an isolated *ray* could be propagated through the ether by means of undulations: still the conception of a ray of light comes so naturally, and has been found so serviceable from the very earliest times, that rays, rather than waves, will be used throughout the present Tract.

Having regard to the apparent rectilinearity of propagation within a homogeneous medium, we may reasonably assume that, if light is propagated by the disturbances of a medium and the disturbances at all parts of

---

[1] Stokes: *Proc. Roy. Soc.*, 1872, vol. 20, p. 442; *Comp. Rend.*, 1873, vol. 77, p. 1150. Abria: *Ibid.*, p. 814. Glazebrook: *Phil. Trans.*, 1880, vol. 171, part 2, p. 421. Hastings: *Amer. J. Sc.*, 1888, ser. 3, vol. 35, p. 60.

the first surface are allowed to produce their effects at the second surface, the resultant disturbance of the area AB is identical with that which would directly follow from a rectilinear transference of the disturbances at points on the area ab to corresponding points on the area AB; and thus that the length Or, which represents the distance to which the front of disturbance has travelled in the direction Or in a given interval of time, also represents the velocity of transmission of a ray of light in the same direction. Regarded from this point of view, the surface of disturbance or wave-surface may be termed the *ray-surface*.

Ray-front.

Further, if a pencil of rays having OrR for axis starts simultaneously from O, the front of the pencil at a certain epoch is a portion of the ray-surface containing r, and at a subsequent epoch is a portion of the ray-

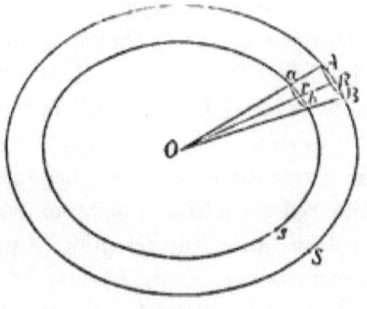

FIG. 2.

surface containing R: in the limiting case, where the pencil is of extremely small angle, its front is in the tangent planes at r and R at successive epochs. The plane front, which thus belongs to an extremely small pencil including a given ray, may be briefly denoted as the *ray-front* for that ray.

Since the ray-surface retains a constant similarity of form and position, for the ratio $OR : Or$ depends solely on the time, the tangent planes at R and r are parallel.

When the ray-surface is not a sphere, the tangent plane at any point is in general inclined obliquely to the radius vector drawn to the point from the origin, and a ray-front is then oblique to its corresponding ray.

In the case of calcite, the ray-surface has two sheets and **consists of a sphere and a spheroid.**

Huygens was thus led to the discovery that the laws of refraction in

the case of calcite are consistent with an undulatory theory, if the velocities of transmission of rays of light within this mineral are determined by a sphere and a spheroid, touching each other in the axis of revolution of the latter.

If a line Or_2r_1 (Fig. 3), drawn from the common centre O, intersects the sphere and spheroid in r_2 and r_1 respectively, according to Huygens the velocity of transmission of one ray in the direction Or_2r_1 is measured by Or_2, and of the other by Or_1.

For a single direction of Or_2r_1, namely that of the axis of revolution $\overline{COC'}$, the two points r_2 and r_1 coincide, and the rays travel with equal velocity; this direction is called the *optic axis* of the crystal.

Plane-polarisation of each of the refracted rays.

So far we have had regard merely to the relation of the two velocities to the direction of ray-transmission within the calcite-crystal. It did not escape the notice of Huygens, however, that each of the rays emergent from a crystal of calcite differs from common light: the difference is one which he was unable to account for by his version of the undulatory theory. It was not till more than a century afterwards (1808) that Malus made the accidental discovery that the same change in the character of the light may be induced by reflection from a plate of glass: to this alteration, termed by Malus *polarisation*, the attention of physicists was largely directed during the immediately succeeding years.

If a plate of calcite be placed with its faces perpendicular to an aerially transmitted plane-polarised ray, the latter is in general divided at the first surface into two: the two rays travel through the plate in directions mutually inclined to each other, and, emerging from it, are transmitted through the air with the same direction as that of the original ray: each of the emergent rays is plane-polarised, but the planes of polarisation of the rays have not the same direction; they are, in fact, perpendicular to each other. Further, when the plate is turned round its normal through any angle, the plane of polarisation of each emergent ray is also displaced through exactly the same angle: the direction of the plane of polarisation is thus dependent on characters belonging to the plate itself: it is found to be independent of the direction of the plane of polarisation of the ray incident on the plate.

For one of the emergent rays, the line of transmission within the plate is continuous with the path of the ray before incidence and after emergence: as the position of the plane of polarisation of the emergent ray is independent of the thickness of the plate, we may reasonably assume, as before in the case of glass, that the ray transmitted *within the plate* is likewise plane-polarised, and that the plane of polarisation during such transmission is identical in direction with that of the *emergent* ray, thus rotating with the plate as the plate is turned round its normal.

As the emergent rays are indistinguishable from each other in character and only differ in the positions in space of their planes of polarisation, we may likewise assume that the second emergent ray is also transmitted within the plate as a plane-polarised ray, but with a direction of plane of polarisation perpendicular to that of the first.

It will be found that the characters of rays which have been transmitted through a plate of calcite can be accounted for, if we imagine that in such a medium a plane-polarised ray transmissible in any given direction has its plane of polarisation in one or other of two rectangular positions, which depend on the crystal itself.

As in the case of air, glass, and analogous media, the periodic change of the ether at every point of a plane-polarised ray transmitted within any bi-refractive medium may be assumed to be dissimilarly symmetrical to two perpendicular planes; but it may be remarked that it is only the *disturbed* ether which is assumed to be dissimilarly symmetrical in the distribution of its characters.[1]

The plane of polarisation is related to the radius vector of the ray-surface.

Malus[2] discovered that the direction of the plane of polarisation of any ray transmitted within a crystal of calcite is determined by the direction of the corresponding radius vector of the ray-surface: he showed that the plane of polarisation for the ray Or_1 (Fig. 3) corresponding to the spheroid is always perpendicular to the plane Or_1C, which contains the ray-direction and the optic axis, and for the ray Or_2 corresponding to the sphere is the plane Or_2C, which also contains the ray-direction and the optic axis: in other words, the plane containing the ray-direction and the optic axis is the plane of polarisation of the ray belonging to the sphere and the transverse plane of the ray belonging to the spheroid.

The above might have led to the recognition of the possible existence and the optical characters of biaxal crystals.

The above facts and reasoning were known to physicists before the existence of biaxal crystals had been discovered; further, the reasoning is really independent of the physical nature of the vibratory change which constitutes light. We proceed to prove that though the geometrical

---

[1] See also pages 7, 79.
[2] *Mém. prés. à l'Institut*: Paris, 1811, vol. 2, p. 413

representation of the laws of transmission of light in biaxal crystals was suggested to Fresnel by ideas in which elasticity had a great part, the possible existence of such crystals, and the corresponding laws of transmission of light, might have been deduced from the above by a simple generalisation, involving no reference either to the constitution of the luminiferous ether or to the nature of the physical change involved in the transmission of light; and further, that the step was so natural a one to take that the discovery of the true form of the wave-surface for biaxal crystals could scarcely have been long avoided.

Another mode of geometrically representing the characters of the extra-ordinarily refracted ray, by reference to the same spheroid, naturally presents itself.

Draw OR_1 parallel to r_1V (Fig. 3), the tangent at r_1 to the ellipse in

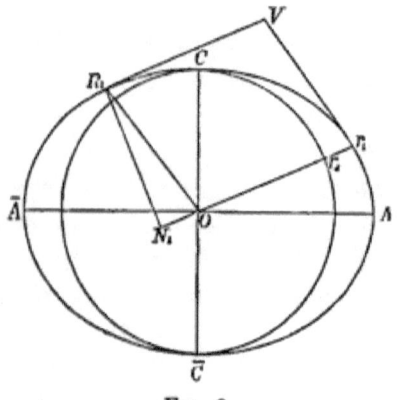

Fig. 3.

which the spheroid is cut by the plane r_1OC; OR_1 and Or_1 are said to be conjugate to each other, and the tangent at R_1 is parallel to Or_1. By a well-known property of the ellipse, the area of the parallelogram of which OR_1, Or_1, are adjacent sides is constant, whatever the direction of Or_1: hence the area is $OA \cdot OC$; OA and OC being the principal axes of the ellipse, and therefore conjugate to each other. But if R_1N_1 is perpendicular to Or_1, meeting it in N_1, and is thus normal to the ellipse and therefore also to the spheroid at R_1, the area of the parallelogram is also $Or_1 \cdot R_1N_1$.

Hence $Or_1 = \dfrac{OA \cdot OC}{R_1N_1}$, whatever the direction of Or_1: and $OA \cdot OC$ being a constant quantity, the *velocity* of a ray transmitted in the direction Or_1 may be represented, not only by Or_1, but by the inverse of R_1N_1. But the same line R_1N_1 determines the *plane of polarisation* of the ray Or_1, for as already stated, the line R_1N_1 is normal to that plane. Further, the same line R_1N_1 determines the *direction* of the ray, for the ray passes through O and is perpendicular to R_1N_1.

Hence the direction, velocity and plane of polarisation of the ray Or_1 can all be represented by means of a single corresponding line R_1N_1, which is at once normal to the spheroid and the ray.

This mode of representation naturally presents itself *as soon as the plane of polarisation is indicated by its normal;* in fact, any attempt to represent geometrically the observed facts of the double refraction of calcite almost inevitably leads to it.

The same mode also suffices to represent the characters of the ordinarily refracted ray without necessitating the use of a second surface.

But for any radius vector Or_1 of a spheroid there are always *two* normals of the spheroid which intersect it perpendicularly: one of them has just been indicated, namely R_1N_1; the other is normal to the plane r_1OC, at the centre of the spheroid, and therefore always lies in the equatorial plane. As already stated, the *plane of polarisation* of the ray Or_2 is r_2OC or r_1OC: and the normal of its plane of polarisation thus lies in the equatorial plane, and is normal both to the spheroid and the ray. Further, the intercept made by the ray Or_2 upon this normal of the spheroid is OA, whatever the direction of Or_2r_1: hence, if the same law as before holds for the relation of the velocity to the intercept upon the normal of the spheroid, the velocity of the ray Or_2 is $\dfrac{OA \cdot OC}{OA}$, or OC; and this is exactly the velocity required.

Hence the velocity and plane of polarisation of the ray Or_2 can likewise be represented by means of a corresponding line which is at once normal to the spheroid and the ray: and this line indicates the plane in which the ray having these characters will lie.

The characters of the refracted rays can be simply expressed by reference to the spheroid alone.

All the characters of rays transmitted in various directions through a crystal of calcite may thus be simply expressed by means of a *single* surface, the spheroid. The relation of the optical characters of the crystal to the geometrical characters of the spheroid is as follows:—

To every given point on a single surface, a spheroid, there in general corresponds one ray: the *direction* of the ray is that of a diameter intersecting perpendicularly the normal drawn at the point to the spheroid; the *velocity* of the ray is inversely proportional to the length of the normal intercepted between the surface and the ray; the *plane of polarisation* of the ray is perpendicular to the same normal.

For points of the spheroid lying on the equatorial circle or at the ends of the axis of revolution, the normal passes through the centre, and the direction of the ray becomes indeterminate: if such a point be regarded as the limiting case of a small ring, it corresponds, not to a single ray, but to an infinity of rays lying in a plane perpendicular to the normal, all transmitted with the same velocity, and all having the same plane of polarisation.

In the case of singly refractive substances there is a spherical surface of reference for which the same general relations are true.

Generalisation.

But it immediately suggests itself that in the case of a crystal like barytes, of which the morphological development and the physical characters are dissimilarly symmetrical to three rectangular planes, the surface of reference, if such a surface exists, is more likely to be an ellipsoid with three unequal axes than an ellipsoid of which two axes are equal. In fact, the correspondence of the optical and the morphological symmetry of crystals was announced by Brewster in 1819.

In the fourth Chapter are deduced the laws of transmission of light in a crystal for which the surface of reference is an ellipsoid having three unequal axes; starting with the hypothesis that the relations between the geometrical characters of the surface of reference and the optical characters of the medium are identical with those which have just been found to obtain when the surface of reference is either a spheroid or a sphere.

The Optical Indicatrix.

To the surface of reference the term *Optical Indicatrix* may be assigned: this suggestive term has the advantage of being equally applicable whether the surface of reference is an ellipsoid, a spheroid, or a sphere, and it is independent of all versions of the undulatory theory; the adjectival prefix may be omitted when the term Indicatrix involves no ambiguity. The Indicatrix is identical in form with the *ellipsoid of elasticity* of various authors, the *ellipsoid of polarisation* of Cauchy, the *ellipsoid of indices* of Mac Cullagh, and the *index-ellipsoid* of Liebisch.

Relation of the optical indicatrix to the general symmetry of the crystal.

In regard to the arrangement of its faces, every crystal is found to belong to one or other of six types of symmetry, distinguished as cubic, tetragonal, hexagonal, ortho-rhombic, mono-symmetric, and anorthic: further, it has been demonstrated by the mathematician that the types of crystalline symmetry thus met with are precisely those which are presented by systems of planes of which the relative positions can be expressed by means of whole numbers, a law to which the faces of crystals are found to conform. Further, we are led by experiment to the induction that a type of symmetry is such, not only for the arrangement of the faces of a crystal, but for all the physical characters: the planes of symmetry characteristic of the types are thus planes of *general symmetry*.

On the other hand, a plane may be one of symmetry for a particular character without being a plane of general symmetry of the crystal: the type is thus not necessarily determinable from the symmetry of the crystal with respect to a single character. For example, a crystal may have the six faces of a cube and really belong, not to the cubic, but to the tetragonal, or even the ortho-rhombic type; observation of some character other than the geometrical being thus necessary to the distinction: again, a plane inclined at any angles to the planes of general symmetry of a cubic crystal, and any plane containing the morphological axis of a tetragonal or hexagonal crystal, is a plane of symmetry for the changes produced by dilatation on change of temperature, and is generally not a plane of symmetry for the facial arrangement.

The above induction requires a plane of general symmetry to be a plane of symmetry of every indicatrix: on the other hand, a plane of symmetry of a single indicatrix is not necessarily a plane of general symmetry of the crystal.

Hence, if the most general form of the indicatrix be an ellipsoid, it will follow that in the case of an ortho-rhombic crystal the axes of any indicatrix must coincide with the three axes of general symmetry. For a tetragonal or hexagonal crystal, the symmetry of the indicatrix with respect to the general planes of symmetry requires two of the axes of the ellipsoid to be equal, and the ellipsoid to be one of revolution about the morphological axis. For a cubic crystal, the symmetry of the indicatrix with respect to the general planes of symmetry necessitates the equality of all the axes of the ellipsoid, and the surface becomes a sphere.

The above is true for all colours of the light, though the relative magnitudes of the axes, both for the general ellipsoid and the ellipsoid of revolution, may vary with the colour: further, it is true for all temperatures of the crystal consistent with the stability of the structure, for a plane of general symmetry must retain that character between the assumed limits of temperature.

In the case of a mono-symmetric crystal, the induction still requires the plane of general symmetry to be a plane of symmetry of the indicatrix for all colours of light and for all temperatures consistent with crystalline stability; but the positions and dimensions of the two axes of the ellipsoid lying in the plane of general symmetry are otherwise independent of the latter, and will in general vary both with the colour of the light and the temperature of the crystal.

And in the case of an anorthic crystal, in which there is a centre, but no plane, of general symmetry, the positions and dimensions of all three rectangular axes of the indicatrix corresponding to a given colour or temperature are free from limitations by a plane of general symmetry, and will likewise vary both with the colour of the light and the temperature at which the determinations are made.

CHAPTER III.

Naturalness of the Method.

Objections.

To the above reasoning, by which it is sought to prove that in the case of calcite the reference of the two sheets to the spheroid alone is one which it is natural to make, and not a mere geometrical artifice only to be discovered after the truth of the generalisation has been established, it may be objected that the reference would in such case have been made long before the present century. It must be remembered, however, that the consequent generalisation would have been a barren speculation at a time when the polarisation of light by reflection was still undiscovered (1808), and the optical characters of most doubly refractive crystals were still beyond the powers of observation; indeed, it was not till the decade 1810-20 that any series of numerical data were available for the testing of a theory: even the accuracy of the construction given by Huygens for the determination of the directions of the refracted rays in calcite was discredited by most physicists at the beginning of this century.

But it may be fairly objected that if the above reference and generalisation were natural, the discovery of the process would have preceded the development of any elastic theory of double refraction. And, in fact, it was really by a process of generalisation that Fresnel's discovery of the true form of the ray-surface for biaxal crystals was made. When the above argument was written, the detailed history of Fresnel's theory had not come to the notice of the author: as the facts are not generally known, and have an important bearing on the true significance of the elastic theory of double refraction, it becomes desirable to explain the position.

The development of Fresnel's theory.

Fresnel's celebrated memoir on Double Refraction was not printed till 1827: in that year, and before the issue of the memoir, Fresnel died at the early age of 39, after years of illness. In the memoir are incorporated papers submitted to the Academy at different dates in the years 1821 and 1822, and it occupies no less than 132 pages of large size. For the sake of brevity, Fresnel made many omissions from the papers as originally submitted to the Academy, and for the sake of clearness adopted a synthetic mode of treatment: the result is that the memoir as printed gives no clue to the real order of discovery, and the reader is apt to infer that Fresnel discovered the true form of the ray-surface *a priori* by means of equations relative to the elastic forces evoked by the disturbance of an incompressible elastic ether. The following statement by Aldis[1] exemplifies this, which is still a very general impression:—

"Fresnel's theory is undoubtedly not a sound dynamical theory. It has, however, the great merit of representing accurately the facts of double refraction as far as experiment at present has tested them, and in one instance has led to the discovery of facts (the conical refractions) previously unobserved. Probably, when the Newton of Physical Optics has succeeded in linking together all the phenomena of Light into one continuous chain, the name of Fresnel will yet be remembered with a reverence akin to that which astronomers feel for Copernicus and Kepler."

The real order of development was of course known to some of Fresnel's

---

[1] W. S. Aldis. *A Chapter on Fresnel's Theory of Double Refraction.* Cambridge, 1870, p. 26.

contemporaries, but to the next generation it was a mystery; it was not till forty years after Fresnel's death that the mystery was dispelled by the publication of the original memoirs, which had been carefully preserved in the family. Verdet,[1] one of the editors of Fresnel's collected papers, makes the following remarks:—

"It may seem odd that reasoning which is incomplete and inexact in two points should have for result one of the best confirmed of the Laws of Nature. But we have seen that this law became manifest to Fresnel as the result of a generalisation quite similar to the generalisations which have led to most great discoveries. When he wished afterwards to account for the law by a mechanical theory, it is not astonishing that he should have led the theory, perhaps unwittingly, towards the end which he already knew of, and that, in his choice of hypotheses, he should have been determined, less by their intrinsic probability, than by their agreement with what he was justified in believing to be true. We have seen some traces of the progress of his ideas in the marginal notes which he had added to the manuscript of memoir No. 38, a memoir here printed for the first time. In the later memoirs we find nothing but the explanation, in different forms, of the mechanical theory by which he tried to demonstrate *a posteriori* the laws which direct intuition had revealed to him."

After this clear statement on the part of his editor, it is obvious that Fresnel's theory of double refraction, however ingenious, has no claim to credit for its predictions: the latter are really a direct consequence of the generalisation which had preceded the theoretical development of the vibratory properties of an elastic but incompressible ether.

Preliminary attempts at generalisation.

The first attempt at the generalisation of Huygens's construction had suggested a sphere combined with a concentric ellipsoid having three unequal axes as the most general form of ray-surface: this assumed that in the most general case one of the rays obeys the ordinary laws of refraction.

It was found, however, that the refraction of the second ray as experimentally determined is inconsistent with an ellipsoidal form of ray-surface. Nor would such a combination of ray-surfaces account for the optical characters of a biaxal crystal: for if a concentric sphere and ellipsoid meet each other, they must either touch at the extremities of a principal diameter, or intersect in two curves; in the former case there would be only one direction of equal ray-velocity; in the latter case this

---

[1] *Œuvres Complètes d'A. Fresnel:* Paris, 1868, vol. 2, p. 327.

character would belong to every diameter which passes through the curves of intersection, and thus to an infinity of lines lying on the surface of a cone.

In 1819 Biot made the important discovery that the results of optical measurement are consistent with two empirical laws, both of them reached by processes of generalisation: in combination with the assumption that one of the rays obeys the laws of ordinary refraction, they completely express the polarisation and velocity of the second ray in terms of its direction in the crystal.[1]

1st law. We have seen that in the case of a uniaxal crystal, two rays transmitted along any given direction had been shown by Malus to have their planes of polarisation respectively coincident and at right angles with the plane containing the ray-direction and the optic axis: from this Biot was led by generalisation to the discovery that in the case of a biaxal crystal the planes of polarisation are the internal and external bisectors of the angle between the two planes which contain the ray-direction and pass each through one of the optic axes.

2nd law. In the case of a uniaxal crystal, if r_1 and r_2 be the velocities of transmission of the two rays transmissible in a direction inclined at an angle σ to the optic axis, it follows from Huygens's construction that the ratio $\left(\dfrac{1}{r_1^2} - \dfrac{1}{r_2^2}\right) : \sin^2\sigma$ is constant for all directions: noticing this, Biot was led by generalisation to the discovery that in the case of a biaxal crystal the ratio $\left(\dfrac{1}{r_1^2} - \dfrac{1}{r_2^2}\right) : \sin\sigma_1 \sin\sigma_2$ is constant; σ_1, σ_2, being the inclinations of the ray to the optic axes.

The second law, combined with the assumption that the velocity of one of the rays is independent of its direction, leads to a surface of the fourth degree, tangent to a concentric sphere at the ends of two diameters, as the ray-surface corresponding to the second ray.

For, let $l\ 0\ n$, $\bar{l}\ 0\ n$, $\lambda\ \mu\ \nu$, be direction-cosines of the optic axes and of the second ray respectively (Fig. 9):

then $\cos\sigma_1 = l\lambda + n\nu$, and $\cos\sigma_2 = -l\lambda + n\nu$.

If r be the variable velocity of the second ray, and a be the constant velocity of the first ray, it follows from the above law that

$$\frac{1}{r^2} - \frac{1}{a^2} = k \sin\sigma_1 \sin\sigma_2,$$

where k is a constant.

---

[1] *Mémoires de l'Acad. de l'Institut de France*, 1820, vol. 3, pp. 228, 233.

Hence $\dfrac{1}{k^2}\left(\dfrac{1}{r^2}-\dfrac{1}{a^2}\right)^2 = (1+\cos\sigma_1)(1-\cos\sigma_1)(1+\cos\sigma_2)(1-\cos\sigma_2)$.

Substituting the values of $\cos\sigma_1$ and $\cos\sigma_2$, and writing

$$\lambda = \dfrac{x}{r},\ \mu = \dfrac{y}{r},\ \nu = \dfrac{z}{r},$$ we have for the equation of the second ray-surface,

$$\dfrac{1}{k^2}\left(1-\dfrac{r^2}{a^2}\right)^2 = (lx+nz+r)(lx+nz-r)(lx-nz+r)(lx-nz-r),$$

an equation of the fourth degree. The second ray-surface meets the first ray-surface ($r = a$) at the intersection of the latter with the four planes, $lx \pm nz \pm a = 0$, and these planes are tangent to the sphere at the extremities of the optic axes: hence the two surfaces are tangent to each other at the same points.

The history of the ray-surface.

At the advent of Fresnel, the emissive theory of light still held almost undisputed sway in the scientific world, notwithstanding the interference discoveries which had been made by Dr. Thomas Young, many years before. Convinced that the true explanation of interference was furnished by the undulatory theory, Fresnel devoted himself with ardour to its theoretical and experimental development, in which he had to sustain the attacks of Laplace, Poisson and Biot, who were firm believers in the truth of the older theory. After successful explanation of diffraction and of the polarisation-colours of thin plates on the undulatory hypothesis, Fresnel in 1821 attempted to solve the problem of double refraction.

Young had already suggested (12 Jan. 1817) that the vibrations of the luminiferous ether are motions transverse to the ray, and had compared the transmission of a ray of light to the transmission of a transverse vibration along a stretched cord:[1] Fresnel himself, in conjunction with Arago, had shown experimentally that two rays polarised in perpendicular planes are incapable of mutual interference, thus confirming the idea that the motion of the ether of a plane-polarised ray is wholly perpendicular to the direction of transmission, so long, at least, as the medium is singly refractive.

Fresnel therefore suggested that the difference of velocities of the two rays transmissible in the same direction in a doubly refractive medium may be due to differences of elastic force evoked by equal displacements of ethereal particles in different directions: the same suggestion had been

---

[1] *Œuvres Complètes d'A. Fresnel*: vol. 1, p. 634; Vol. 2, p. 742.

published by Young[1] in 1809, before the idea of transverse vibrations had presented itself. Assuming that the ethereal elasticity, in the case of a uniaxal crystal, is the same for all directions of displacement perpendicular to the optic axis, and different from that evoked by displacements parallel to it, and that the vibrations are always *perpendicularly* transverse to the ray[2] (an assumption he departed from later), Fresnel showed that the vibration of the ordinary ray must be perpendicular to its plane of polarisation; for in this case the vibration, being always perpendicular to the optic axis, evokes an elastic force of which the magnitude is the same whatever the direction of the ray. The vibration of the extraordinary ray being likewise assumed to be perpendicular both to the ray and its plane of polarisation, the elastic force evoked by it will vary with the direction of the vibration, and thus the velocity will depend on the inclination of the ray to the optic axis.

But Fresnel[3] soon saw that, if such an explanation is true, neither of the rays transmitted in a biaxal crystal can have a velocity independent of the direction, for in such a crystal there is no direction such that the optical characters are the same in all directions equally inclined to it. Having submitted this inference to the test of experiment, Fresnel announced its confirmation in September 1821, thus completely upsetting the ideas which then prevailed as to the forms and relations of the two ray-surfaces of a biaxal crystal.

Further developing the theory, Fresnel[4] showed (November 19, 1821) that if the ethereal elasticity be proportional to the square of the velocity, as in the case of the longitudinal vibrations of an elastic medium, a surface such that any radius vector represents the square of the elastic force evoked by a unit-displacement in its direction is, in the case of a uniaxal crystal, a spheroid (distinct from the spheroidal ray-surface itself), *if the double refraction is small*. Conversely, lines measured in a direction perpendicular to a diametral plane of the auxiliary spheroid, and having lengths equal to the maximum and minimum radii vectores of the section, would *approximately* represent the velocities with which vibrations, parallel to those radii vectores, would be transmitted along the normal of the plane.

It then suggested itself to Fresnel that an ellipsoid with three unequal axes might be a more general form of this surface of elasticity, and that the same construction might hold for the determination of the *approximate* velocities of the rays having a given direction. The two circular

---

[1] *History of the Inductive Sciences*; by Whewell: London, 1857, vol. 2, p. 329.
[2] *Œuvres Complètes d'A. Fresnel:* vol. 2, p. 281.
[3] *Ibid.*, p. 257.
[4] *Ibid.*, pp. 283, 304, 306.

sections of the ellipsoid immediately account for the existence of two optic axes, for there will be only one velocity of transmission along their normals: again, the planes of polarisation for a given ray-direction are at right angles, for the corresponding vibrations are parallel to the axes of the diametral section perpendicular to the common direction. It remained to discover whether the two empirical laws established by Biot were geometrically consistent with the form of ray-surface suggested by the above construction: and this being found by Fresnel to be very approximately the case, the true form of ray-surface was at last determined.

In this way, however, it presented itself as an approximation, true only when the double refraction is small: it was not till later (26 Nov. 1821) that the hypothetical form of the surface of elasticity[1] was changed from an ellipsoid to a surface of the fourth degree: with this hypothesis the ray-surface already obtained as an approximation appeared as the true ray-surface, independently of the amount of the double refraction of the medium.

It follows from the above that Fresnel's discovery of the form of the ray-surface for biaxal crystals was really arrived at by a geometrical generalisation of Huygens's ray-surface for uniaxal crystals, and that the geometrical relation used in the generalisation was suggested by the conception of a plane-polarised ray as due to vibratory motion perpendicular to its direction.

The true nature of the luminiferous ether.

Notwithstanding the success with which so many optical properties have been explained on the hypothesis that light is a vibratory motion of an elastic ether of which the effective density depends on the permeated matter, it would be wrong to infer that light is *actually* due to such a vibratory motion. It is conceivable that other hypotheses may likewise lead to similar results: and, indeed, any other quantities about which the same general assertions may be made and which obey the same mathematical laws will satisfy the equations and furnish other analogies. In fact, Prof. Gibbs[2] has lately shown that the equations which result from the last version of the elastic theory have a corresponding electrical interpretation.

Again, according to the late Prof. Clerk Maxwell, light may be an electro-magnetic disturbance propagated according to electro-magnetic laws; and he showed that the velocity of transmission of electro-magnetic

---

[1] *Œuvres Complètes d'A. Fresnel:* vol. 2, p. 338.
[2] *Phil. Mag.* 1889, ser. 5, vol. 27, p. 238.

disturbance in free ether is identical with that of light. The recent experiments of Dr. Hertz have placed the undulatory character of electromagnetic radiation beyond the region of hypothesis, and it is now experimentally established that electro-magnetic waves and light-waves differ only in length; while the wave-length for sodium-light is ·000589 millimetres, the short electro-magnetic waves produced by Dr. Hertz had still a length of 2 metres. On the other hand, it seems that electro-magnetic actions are inexplicable as mere vibratory motion of an elastic ether: Prof. Fitzgerald,[1] who has given much thought to the mechanical representation of the ether, points out that "if magnetic forces are analogous to the rotation of the elements of a wave, an ordinary solid cannot be analogous to the ether, because the latter may have a constant magnetic force existing in it for any length of time, while an elastic solid cannot have continuous rotation of its elements in one direction existing within it."

The educational difficulty.

Taking everything into consideration, it seems undesirable, from the purely educational point of view, to continue such a synthetic mode of treatment as was adopted by Fresnel in the memoir of 1827. The subject of the optical properties of crystals is so extensive that it is unsatisfactory to make all the laws appear to depend upon an hypothesis of the truth of which we are not convinced: otherwise it becomes necessary either to keep the student in ignorance of the doubts as to the truth of the hypothesis, or to raise a feeling of distrust as to the accuracy of every deduction therefrom. It would seem better to develop the subject by means of analogy and experiment, and to assign a subordinate importance to the mechanism of the ether.

Three other modes of generalisation.

In addition to the method explained in the last Chapter, there are at least three others by which a generalisation of Huygens's construction may be arrived at. One of them depends on the fact that the wave-surface is an envelope.

If Of_2f_1 (Fig. 4) be normal to parallel tangent planes of the sphere and spheroid, and we consider a section of the surface by a plane containing the optic axis and the given line, it follows that $Of_1 = \dfrac{AO \cdot OC}{OR_1}$, where OR_1 is per-

---

[1] *Nature*, 1890, vol. 42, p. 173.

pendicular to Of_1, and that Of_2 is equal to OC or $\dfrac{OA \cdot OC}{OA}$. Hence the lines Of_1, Of_2, which represent the velocities of the wave-fronts f_1r_1, f_2g, are inversely proportional to the lines OR_1 and OA, or to the axes of the ellipse in which the spheroid is intersected by a plane parallel to the wave-fronts $f_1 r_1$, $f_2 g$.

Huygens's wave-surface is thus the envelope of the two planes which are parallel to the same central section of the spheroid, and pass through those points on the common central normal which are distant from the centre by lengths inversely proportional to the axes of the section. On generalising this result, by the substitution of an ellipsoid with three unequal axes for the spheroid, Fresnel's wave-surface is obtained:

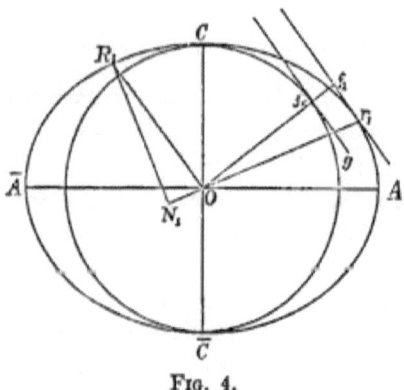

Fig. 4.

this generalised construction is virtually identical with the one employed by Fresnel in the memoir of 1827; the only difference being that he used the inverse of the ellipsoid instead of the ellipsoid itself.

The two other geometrical constructions, which present themselves for the generalisation of Huygens's construction, only appear indirectly: they depend on corresponding relations between the rays or ray-fronts and the *polar reciprocal of the indicatrix*.

Advantages of the method here suggested.

There are thus two purely geometrical processes, which *directly* present themselves for trial as being possible modes of representing the optical characters of those crystals which belong to a lower type of general symmetry than the uniaxal: both lead to identical results. If we are compelled to select one or other of these processes for use as an educational instrument, there can be no doubt on which the choice must fall.

The method based on an envelope is so far wanting in simplicity that Fresnel himself gave no rigorous solution of the equations: this was supplied by Ampère, and it was not till even a decade later that the less complicated mode of elimination, now generally given, was invented by Archibald Smith. Further, the construction yields the wave-surface in such a way that its singularities are not obvious, and were only remarked by Sir William Hamilton several years after Fresnel's death.

On the other hand, as was shown in Chapter II, the geometrical basis here advocated naturally suggests itself as soon as any attempt is made to represent geometrically the observed optical properties of uniaxal crystals: we shall further show that it readily furnishes the equation of the ray-surface without demanding any knowledge of the differential calculus or any determination of maxima and minima; that it immediately suggests all the singularities of the ray-surface; and that, in fact, most optical problems are reduced to a form in which their solution can be effected by the elementary geometry of the ellipsoid. We may add that the employment of an additional ellipsoid, the polar reciprocal of the first, is rendered unnecessary, and a continual source of confusion to the student is thereby removed.

CHAPTER IV.

Deduction of the Optical Characters corresponding to an Ellipsoidal Indicatrix.

General Relation.

The characters of a ray of plane-polarised homogeneous light transmitted within a medium are indicated by geometrical characters at a corresponding point on an ellipsoid; the direction of the ray is that of a diameter intersecting perpendicularly the normal drawn to the ellipsoid at the corresponding point: the velocity is inversely proportional to the length of the normal intercepted by the ray; the plane of polarisation is perpendicular to the normal.

It is required to deduce the relations of the optical characters for different directions in the medium.

The order of deduction is as follows:—

Arts. 1-15 relate to *rays* in general; Arts. 16-20, to the pair of ray-directions here termed *bi-radials;* Arts. 21-40, to *ray-fronts* in general; Arts. 41-44, to the pair of front-normals here termed *bi-normals;* Arts. 45-52, to the *bi-radial* and *bi-normal cones.*

1. The construction of the ray-surface.

Let $a^2x^2 + b^2y^2 + c^2z^2 = 1$ be the equation of the indicatrix; a, b, c being in descending order of magnitude: O the centre of the indicatrix: $x'\ y'\ z'$ the co-ordinates of R, a point on the indicatrix: NOr a line intersecting the normal RN perpendicularly (Fig. 5).

According to the above relation, the *direction of the ray* corresponding to the point R of the indicatrix is given by the line NOr, the *velocity* of the ray is measured by $\dfrac{1}{RN}$: the *plane of polarisation* is perpendicular to RN.

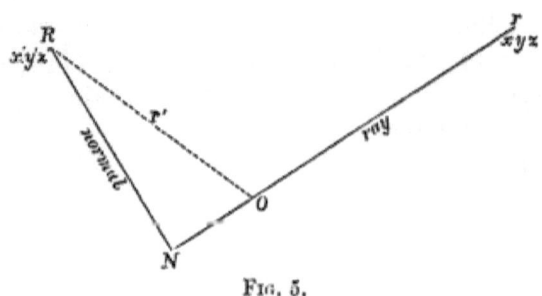

Fig. 5.

Take the length Or equal to $\dfrac{1}{RN}$: the locus of the points r, corresponding to all positions of R on the indicatrix, will be the *ray-surface* of the medium for the given simple colour: since the velocity of transmission of a ray of light of that colour along any radius vector of the surface is measured by the length of the radius vector.

The propositions of the present Chapter are stated in a form which is independent of any particular version of the undulatory theory: it may be remarked, however, that according to the latest version of the elastic theory, RN is the *direction of vibration* for the ray Or; according to Fresnel's version of the elastic theory and the present statement of the electro-magnetic theory, the direction of vibration is RO.[1]

---

[1] *Philos. Magazine,* 1888, ser. 5, vol. 26, p. 528: *Electricity and Magnetism,* by J. Clerk Maxwell; Oxford, 1881, vol. 2, p. 404; *Nature,* 1890, vol. 42, p. 174.

2. The symmetry of the ray-surface.

From the mode of construction, it is evident that the planes of symmetry of the indicatrix are also planes of symmetry of the ray-surface.

3. The sections of the ray-surface by the planes of symmetry.

The section of the ray-surface by each plane of symmetry consists of a circle and an ellipse; the radius of the circle is the inverse of that axis of the indicatrix which is perpendicular to the plane of the section; the ellipse is similar and similarly situated to the section of the indicatrix by the same plane. This may be proved as follows:—

Let $AO\bar{A}$, $BO\bar{B}$, $CO\bar{C}$, be the principal axes of the indicatrix, and $OA = \frac{1}{a}$, $OB = \frac{1}{b}$, $OC = \frac{1}{c}$: it is required to determine the section of the ray-surface by one of the planes of symmetry, say AOC (Fig. 6).

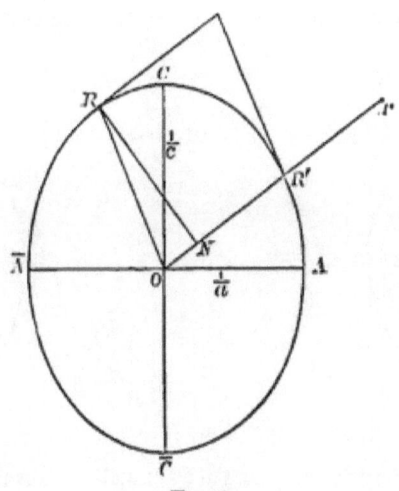

Fig. 6.

a. By considering a series of points in a small ring surrounding the point B on the indicatrix, it is seen that in the limit the point B itself corresponds, not to one ray, but to an infinity of rays, all lying in the plane AOC; for the axis OB is the normal of the indicatrix at B, and intersects perpendicularly all radii vectores of the indicatrix which lie in the plane AOC. Further, the length of the normal at B intercepted by each of the rays is OB: hence the velocity of each ray corresponding to the point B is $\frac{1}{OB}$ or b, and a circle of radius b, situated in the plane AOC, is on the ray-surface (Fig 7c).

D

b. For a point R on the indicatrix, lying in the plane AOC, the normal RN of the indicatrix lies in the plane AOC, and is also normal at the point R to the elliptic section $AC\bar{A}\bar{C}$: if ONr is perpendicular to RN and $Or = \dfrac{1}{RN}$, Or is by construction a radius vector of the ray-surface (Fig. 6).

If R' be a point in which Or intersects the indicatrix, OR and OR' are conjugate to each other, for Or being perpendicular to RN is parallel to the tangent of the ellipse at R: hence the product $OR' \cdot RN$, which measures the area of the parallelogram of which OR, OR', are adjacent sides, is constant, and equals the product $OA \cdot OC$ or $\dfrac{1}{ac}$.

Hence, $Or = ac \, OR'$.

The locus of r is thus an ellipse, similar and similarly situated to the ellipse $AC\bar{A}\bar{C}$, and having ac times its linear dimensions: its semi-axis in the line OA will thus be c, and its semi-axis in the line OC will be a (Fig. 7c).

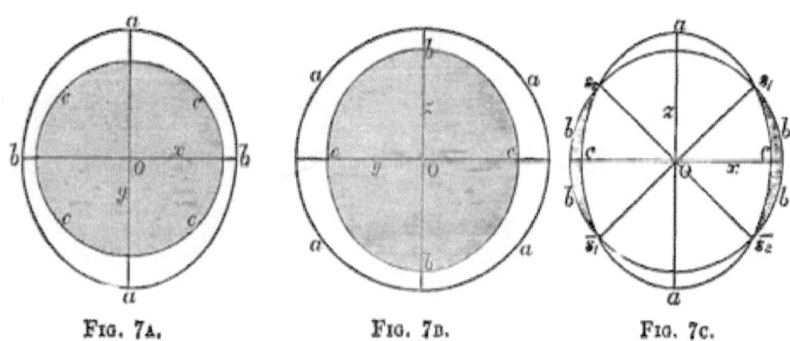

FIG. 7A.   FIG. 7B.   FIG. 7C.

c. To any ray lying in the plane AOC no other normal of the indicatrix is perpendicular: hence no radius vector of the ray-surface, other than the above, lies in the plane AOC; the circle $z^2 + x^2 = b^2$ and the ellipse $\dfrac{z^2}{a^2} + \dfrac{x^2}{c^2} = 1$ are thus the only curves of intersection of the ray-surface with the plane AOC (Fig. 7c).

d. Similarly, the section of the ray-surface by the plane of symmetry BOA is a circle $x^2 + y^2 = c^2$ and an ellipse $\dfrac{x^2}{b^2} + \dfrac{y^2}{a^2} = 1$ (Fig. 7A).

e. And the section of the ray-surface by the plane of symmetry COB is a circle $y^2 + z^2 = a^2$, and an ellipse $\dfrac{y^2}{c^2} + \dfrac{z^2}{b^2} = 1$ (Fig. 7B).

f. On consideration of the relative magnitudes of a, b, c, it will be seen that in one of the planes of symmetry the circle falls wholly within the ellipse; in another, the ellipse falls wholly within the circle; in the third, containing the longest and shortest axes of the indicatrix, the circle and ellipse intersect each other in four points lying at the extremities of two diameters $s_1 \bar{s}_1, s_2 \bar{s}_2$.

g. For any direction of ray lying in a plane of symmetry, there are thus two possible directions for the plane of polarisation, perpendicular to each other; and in general each plane of polarisation corresponds to a different velocity of transmission in the given direction. For two directions of the ray, those of the diameters $s_1 \bar{s}_1, s_2 \bar{s}_2$, the two velocities are equal (Fig. 7c).

If $x\, 0\, z$ be the co-ordinates of one of the points s, we have
$$\frac{x^2}{c^2} + \frac{z^2}{a^2} = 1 \text{ and } x^2 + z^2 = b^2 : \text{ hence}$$
$$\frac{x^2}{c^2(a^2 - b^2)} = \frac{z^2}{a^2(b^2 - c^2)}.$$

If $\lambda\, 0\, \nu$ be the direction-cosines of a diameter Os, the relation may also be written as
$$\frac{a^2 \lambda^2}{b^2 - a^2} + \frac{c^2 \nu^2}{b^2 - c^2} = 0.$$

The angle $s_1 OC$ is given by the relation
$$\tan s_1 OC = \frac{c\sqrt{(a^2 - b^2)}}{a\sqrt{(b^2 - c^2)}} = \frac{\sqrt{\left(\frac{1}{a^2} - \frac{1}{b^2}\right)}}{\sqrt{\left(\frac{1}{b^2} - \frac{1}{c^2}\right)}}.$$

h. In each of the planes OBC, OCA, OAB, a plane-polarised ray is thus transmissible with the velocity $a, b,$ or c, respectively, whatever its direction in the plane: hence, by the general principle of undulations, the refraction of these rays by a surface perpendicular to the symmetral plane will be ordinary, and the index of refraction will be $\frac{v}{a}, \frac{v}{b}, \frac{v}{c}$, respectively; v being the constant velocity of transmission in the other medium. If a, β, γ, be the values of the index of refraction of the rays in each of the above planes, for which the index of refraction is independent of the direction, we must have
$$a : b : c = \frac{1}{a} : \frac{1}{\beta} : \frac{1}{\gamma};$$

a, β, γ, are termed the *principal indices of refraction*.

4. *Given the co-ordinates $x'y'z'$ of R, to find the velocity r of the corresponding ray Or.*

The velocity of the ray Or is measured by $\frac{1}{RN}$ (Fig. 5): but RN, being the normal of the indicatrix at the point R and perpendicular to ON, by construction is equal to the perpendicular drawn from the origin O to the plane $a^2x'x + b^2y'y + c^2z'z = 1$ which touches the indicatrix at R. If p be the length of this perpendicular

$$\frac{1}{p} = \sqrt{(a^4x'^2 + b^4y'^2 + c^4z'^2)}.$$

Hence $\quad r^2 = \dfrac{1}{RN^2} = \dfrac{1}{p^2} = a^4x'^2 + b^4y'^2 + c^4z'^2:$

r being the length of that radius vector of the ray-surface which corresponds to the point R of the indicatrix.

5. *Given the co-ordinates $x'y'z'$ of R, to find the direction-cosines of the normal of the plane of polarisation of the corresponding ray Or.*

The plane of polarisation, being perpendicular to the normal RN (Fig. 5), is parallel to the tangent plane of the indicatrix at the point R; the equation of the tangent plane may be written in the form

$$pa^2x'x + pb^2y'y + pc^2z'z = p.$$

Hence the direction-cosines of the normal of the plane of polarisation are

$$pa^2x',\ pb^2y',\ pc^2z';$$

or $\quad \dfrac{a^2x'}{r},\ \dfrac{b^2y'}{r},\ \dfrac{c^2z'}{r}:$

where $r^2 = a^4x'^2 + b^4y'^2 + c^4z'^2$.

6. *Given the co-ordinates $x'y'z'$ of R, to find the direction-cosines of the corresponding ray Or.*

The direction-cosines of RN (Fig. 5) being $\dfrac{a^2x'}{r},\ \dfrac{b^2y'}{r},\ \dfrac{c^2z'}{r}$, and those of OR being $\dfrac{x'}{r'},\ \dfrac{y'}{r'},\ \dfrac{z'}{r'}$, where $r' = OR$, the direction-cosines $h\,k\,l$ of a line perpendicular to both RN and OR, and therefore to the plane RON and all lines therein, are given by the equations:—

$$hx' + ky' + lz' = 0$$
$$ha^2x' + kb^2y' + lc^2z' = 0:$$

hence

$$\frac{h}{y'z'(b^2-c^2)} = \frac{k}{z'x'(c^2-a^2)} = \frac{l}{x'y'(a^2-b^2)}.$$

If $\lambda\,\mu\,\nu$ be the direction-cosines of any line whatsoever in the plane RON,
$$\lambda h + \mu k + \nu l = 0$$
or $\lambda y'z'(b^2-c^2) + \mu z'x'(c^2-a^2) + \nu x'y'(a^2-b^2) = 0$.

Let the line $\lambda\,\mu\,\nu$ coincide with Or, in which case it is at right angles to RN, of which the direction-cosines are $\dfrac{a^2x'}{r}, \dfrac{b^2y'}{r}, \dfrac{c^2z'}{r}$; we thus have
$$\lambda a^2 x' + \mu b^2 y' + \nu c^2 z' = 0.$$
Determining the ratios $\lambda : \mu : \nu$ from the last two equations, we get
$$\frac{\lambda}{A} = \frac{\mu}{B} = \frac{\nu}{C}$$
where
$$A = x'z'^2 c^2(c^2-a^2) - x'y'^2 b^2(a^2-b^2) = x'(r^2-a^2),$$
$$B = y'x'^2 a^2(a^2-b^2) - y'z'^2 c^2(b^2-c^2) = y'(r^2-b^2),$$
$$C = z'y'^2 b^2(b^2-c^2) - z'x'^2 a^2(c^2-a^2) = z'(r^2-c^2);$$
or $\dfrac{\lambda}{x'(r^2-a^2)} = \dfrac{\mu}{y'(r^2-b^2)} = \dfrac{\nu}{z'(r^2-c^2)}$

in which $r^2 = a^4 x'^2 + b^4 y'^2 + c^4 z'^2$, as before.

These equations determine the direction-cosines $\lambda\,\mu\,\nu$ of the ray Or corresponding to the point $x'y'z'$ or R.

7. The equation of the ray-surface.

The co-ordinates of r being $x\,y\,z$, we have $x = \lambda r, y = \mu r, z = \nu r$: $\lambda\,\mu\,\nu$ being the direction-cosines of the ray Or.

Hence, substituting these values in the last set of equations,
$$\frac{\dfrac{x}{r^2-a^2}}{x'} = \frac{\dfrac{y}{r^2-b^2}}{y'} = \frac{\dfrac{z}{r^2-c^2}}{z'}.$$

Each of these fractions is equal to
$$\frac{a^2 x \dfrac{x}{r^2-a^2} + b^2 y \dfrac{y}{r^2-b^2} + c^2 z \dfrac{z}{r^2-c^2}}{a^2 xx' + b^2 yy' + c^2 zz'}.$$

But the denominator of the last expression is zero; for by construction the line Or, of which the direction-cosines are $\dfrac{x}{r}, \dfrac{y}{r}, \dfrac{z}{r}$, is perpendicular to RN, of which the direction-cosines are $\dfrac{a^2 x'}{r}, \dfrac{b^2 y'}{r}, \dfrac{c^2 z'}{r}$ (Art. 5).

Hence the numerator is also zero; for the fractions equivalent to the expression are never all of them indeterminate, and are never infinite.

Thus $\dfrac{a^2 x^2}{r^2-a^2} + \dfrac{b^2 y^2}{r^2-b^2} + \dfrac{c^2 z^2}{r^2-c^2} = 0:$

this is the equation of the ray-surface, for it expresses a relation between the co-ordinates $x\,y\,z$ of any point r lying in it.

The equation may also be written in the form

$$\frac{a^2 x^2}{r^2-a^2} + x^2 + \frac{b^2 y^2}{r^2-b^2} + y^2 + \frac{c^2 z^2}{r^2-c^2} + z^2 = r^2,$$

or

$$\frac{x^2}{r^2-a^2} + \frac{y^2}{r^2-b^2} + \frac{z^2}{r^2-c^2} = 1.$$

8. *Given* $\lambda\,\mu\,\nu$, *the direction-cosines of a line of transmission, to find* r_1 *and* r_2, *the velocities of the corresponding rays* Or_1, Or_2.

Substituting the values $x=\lambda r$, $y=\mu r$, $z=\nu r$ in the equation of the ray-surface, we have

$$\frac{a^2\lambda^2}{r^2-a^2} + \frac{b^2\mu^2}{r^2-b^2} + \frac{c^2\nu^2}{r^2-c^2} = 0:$$

and, multiplying out,

$$r^4 (a^2\lambda^2 + b^2\mu^2 + c^2\nu^2) - r^2 \{a^2(b^2+c^2)\lambda^2 + b^2(c^2+a^2)\mu^2 + c^2(a^2+b^2)\nu^2\} + a^2 b^2 c^2 = 0.$$

This being a quadratic equation in r^2, there are in general, for given values of $\lambda\,\mu\,\nu$, two solutions, say r_1^2 and r_2^2, and thus two velocities of transmission in the given direction.

A geometrical solution is given in Art. **15**.

The above equation may sometimes be conveniently written in the form

$$\frac{\lambda^2}{\frac{1}{r^2}-\frac{1}{a^2}} + \frac{\mu^2}{\frac{1}{r^2}-\frac{1}{b^2}} + \frac{\nu^2}{\frac{1}{r^2}-\frac{1}{c^2}} = 0.$$

9. *Given* $\lambda\,\mu\,\nu$, *the direction-cosines of a line of transmission, to find the co-ordinates* $x_1'\,y_1'\,z_1'$, $x_2'\,y_2'\,z_2'$ *of the points* R_1, R_2, *of the indicatrix which correspond to the rays* Or_1, Or_2, *respectively.*

Having found r_1^2 and r_2^2, as indicated in the last Article, the co-ordinates $x_1'\,y_1'\,z_1'$, $x_2'\,y_2'\,z_2'$ are given by the equations (Art. 6):—

$$\frac{x_1'}{\frac{\lambda}{r_1^2-a^2}} = \frac{y_1'}{\frac{\mu}{r_1^2-b^2}} = \frac{z_1'}{\frac{\nu}{r_1^2-c^2}} = L_1 \text{ (say)},$$

and

$$\frac{x_2'}{\frac{\lambda}{r_2^2-a^2}} = \frac{y_2'}{\frac{\mu}{r_2^2-b^2}} = \frac{z_2'}{\frac{\nu}{r_2^2-c^2}} = L_2 \text{ (say)}:$$

remembering that

$$a^2 x_1'^2 + b^2 y_1'^2 + c^2 z_1'^2 = 1,$$
$$\text{and } a^2 x_2'^2 + b^2 y_2'^2 + c^2 z_2'^2 = 1,$$

since the points R_1, R_2, are on the indicatrix.

10. *The points R_1, R_2, corresponding to the rays Or_1, Or_2, transmissible along the line $\lambda\mu\nu$, are in a plane conjugate to that line.*

Since the normals of the indicatrix at R_1, R_2, are both perpendicular to the line $\lambda\,\mu\,\nu$ we have
$$\lambda a^2 x_1' + \mu b^2 y_1' + \nu c^2 z_1' = 0$$
$$\lambda a^2 x_2' + \mu b^2 y_2' + \nu c^2 z_2' = 0.$$
Hence the points R_1, R_2, are in the plane
$$\lambda a^2 x + \mu b^2 y + \nu c^2 z = 0.$$
This is the equation of a plane passing through the centre of the indicatrix and parallel to the planes which touch the indicatrix at either of the points where the line $\lambda\,\mu\,\nu$ intersects it.

It is also obvious geometrically that the tangent planes at R_1, R_2, are both of them parallel to Or, and that Or is therefore parallel to their line of intersection; Or is thus conjugate to the plane containing the points O, R_1, R_2.

And it is geometrically evident that at *all* points of the section made by the conjugate plane the tangent planes to the indicatrix are parallel to, and therefore their normals perpendicular to, the conjugate line $\lambda\,\mu\,\nu$: the points R_1, R_2, are those of the section for which the normals of the indicatrix are not only perpendicular to the line $\lambda\,\mu\,\nu$, but intersect it.

11. *Given the direction of transmission, to find the positions of the corresponding points R_1, R_2, in the conjugate plane.*

From the last Article, it follows that the tangent planes to the indicatrix at its intersection with the conjugate plane form a tangent cylinder, having its axis parallel to the direction of transmission. Let UKV be the curve of contact of the cylinder and indicatrix (Fig. 8): R_1, R_2, are somewhere on the curve UKV.

As a line is only perpendicular to its conjugate plane when it coincides with an axis of the indicatrix, Or, the axis of the cylinder, is in general oblique to UKV, the conjugate plane.

Let $U'K'V'$, $U''K''V''$ be sections of the cylinder by two planes perpendicular to its axis; they are in general ellipses: let $K''KK'$ be any line on the cylinder parallel to the axis, and $K''L''$, KL, $K'L'$, be the normals of the cylinder at the points K'', K, K', respectively; they are evidently parallel to each other.

But KL is also the normal of the indicatrix at K, for the cylinder and indicatrix are tangent to each other at that point: also $K'L'$ lies in the plane $U'K'V'$, since that plane is perpendicular to the axis of the cylinder: hence $K'L'$ is the normal of the ellipse $U'K'V'$ at the point K'.

The line KL will thus only intersect the axis of the cylinder when $K''L'$ is an axis of the section $U''K'V'$.

Hence the points R_1, R_2, $\overline{R}_1$, $\overline{R}_2$, are the four positions of K, on the curve UKV, for which the normal of the indicatrix intersects the axis of the cylinder: and these four positions are projections, by lines parallel to the axis of the cylinder, of the extremities of the axes of its "base;" the base being taken as perpendicular to the axis of the cylinder.

In other words, the points R_1, R_2, and the normals R_1N_1, R_2N_2, lie in the *planes of symmetry of that tangent cylinder* of the indicatrix which has its axis in the common direction of transmission of the rays.

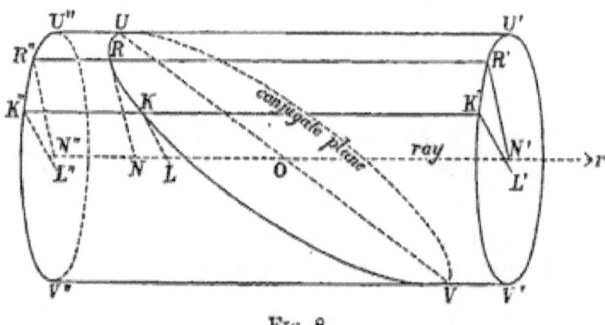

Fig. 8.

12. *The planes of polarisation of the two rays Or_1, Or_2, transmissible along the same line are perpendicular to each other.*

By the last Article, the normals R_1N_1, R_2N_2, are in the planes of symmetry of the tangent cylinder and at right angles to its axis: hence the planes of polarisation, to which the two lines are perpendicular, are themselves perpendicular to each other.

The following analytical proof is interesting by reason of the eliminations:—

The normals of the planes of polarisation being normals of the indicatrix at R_1, R_2, their direction-cosines are

$$\frac{a^2 x_1'}{r_1},\ \frac{b^2 y_1'}{r_1},\ \frac{c^2 z_1'}{r_1},$$

$$\frac{a^2 x_2'}{r_2},\ \frac{b^2 y_2'}{r_2},\ \frac{c^2 z_2'}{r_2},\ \text{respectively (Art. 5);}$$

hence, if ϕ be the angle between the planes of polarisation,

$$\cos \phi = \frac{1}{r_1 r_2}(a^4 x_1' x_2' + b^4 y_1' y_2' + c^4 z_1' z_2').$$

Substituting the values of $x_1'y_1'z_1'$, $x_2'y_2'z_2'$, from Article 9, we have
$$\cos\phi = \frac{L_1 L_2}{r_1 r_2}\left\{\frac{a^4\lambda^2}{(r_1^2-a^2)(r_2^2-a^2)}+\frac{b^4\mu^2}{(r_1^2-b^2)(r_2^2-b^2)}+\frac{c^4\nu^2}{(r_1^2-c^2)(r_2^2-c^2)}\right\}.$$

Now r_1^2, r_2^2, being the roots of the equation given in Article 8, we have
$$r_1^2 + r_2^2 = \frac{a^2(b^2+c^2)\lambda^2 + b^2(c^2+a^2)\mu^2 + c^2(a^2+b^2)\nu^2}{a^2\lambda^2 + b^2\mu^2 + c^2\nu^2}$$

and
$$r_1^2 r_2^2 = \frac{a^2 b^2 c^2}{a^2\lambda^2 + b^2\mu^2 + c^2\nu^2};$$

hence
$$(r_1^2-a^2)(r_2^2-a^2) = r_1^2 r_2^2 - a^2(r_1^2+r_2^2) + a^4$$
$$= -\frac{a^2\lambda^2}{b^2-c^2}\frac{(a^2-b^2)(b^2-c^2)(c^2-a^2)}{a^2\lambda^2+b^2\mu^2+c^2\nu^2}.$$

Similarly, $(r_1^2-b^2)(r_2^2-b^2) = -\dfrac{b^2\mu^2}{c^2-a^2}\dfrac{(a^2-b^2)(b^2-c^2)(c^2-a^2)}{a^2\lambda^2+b^2\mu^2+c^2\nu^2}$,

and $(r_1^2-c^2)(r_2^2-c^2) = -\dfrac{c^2\nu^2}{a^2-b^2}\dfrac{(a^2-b^2)(b^2-c^2)(c^2-a^2)}{a^2\lambda^2+b^2\mu^2+c^2\nu^2}.$

Hence $\dfrac{a^4\lambda^2}{(r_1^2-a^2)(r_2^2-a^2)}+\dfrac{b^4\mu^2}{(r_1^2-b^2)(r_2^2-b^2)}+\dfrac{c^4\nu^2}{(r_1^2-c^2)(r_2^2-c^2)}=0,$

and ϕ is a right angle.

13. *If ρ be the point in which Or intersects the indicatrix, the lines $O\rho$, OR_1, OR_2, form a triad of conjugate diameters of the indicatrix.*

The axes of the basal section $U'K'V'$ (Fig. 8) of the tangent cylinder of the indicatrix being conjugate to each other, it follows, from the properties of parallel projection, that the projections of the axes on any section of the cylinder are conjugate diameters of the curve of section: hence the line $O\rho$ and the lines OR_1, OR_2 (which are the projections of the basal axes on the conjugate plane of Or), form a triad of conjugate diameters of the indicatrix, each being conjugate to the plane of the other two.

This may also be proved analytically, as follows:—

Substituting the values of $x_1' y_1' z_1'$, $x_2' y_2' z_2'$, given in Art. 9, we find that
$$a^2 x_1' x_2' + b^2 y_1' y_2' + c^2 z_1' z_2' =$$
$$L_1 L_2\left\{\frac{a^2\lambda^2}{(r_1^2-a^2)(r_2^2-a^2)}+\frac{b^2\mu^2}{(r_1^2-b^2)(r_2^2-b^2)}+\frac{c^2\nu^2}{(r_1^2-c^2)(r_2^2-c^2)}\right\}.$$

The quantity within the brackets is zero, as may be seen from the last Article, or more directly by subtraction of the equations
$$\frac{a^2\lambda^2}{r_1^2-a^2}+\frac{b^2\mu^2}{r_1^2-b^2}+\frac{c^2\nu^2}{r_1^2-c^2}=0$$
$$\frac{a^2\lambda^2}{r_2^2-a^2}+\frac{b^2\mu^2}{r_2^2-b^2}+\frac{c^2\nu^2}{r_2^2-c^2}=0.$$

Hence $a^2 x_1' x_2' + b^2 y_1' y_2' + c^2 z_1' z_2' = 0$; but this is the condition that the lines OR_1, OR_2, may be parallel to the tangent planes at R_2 and R_1 respectively: as those tangent planes are likewise both parallel to the line Or, the three lines form a triad of conjugate diameters.

Corollary. The normal of the indicatrix at the point R_1, though it intersects the line Or, only intersects the line, OR_2 when N_1 coincides with O (Fig. 14); hence it follows that if OR_2 be a line of ray-transmission, R_1 is *not* one of the corresponding points on the indicatrix: in general the lines of the conjugate triad $O\rho$, OR_1, OR_2, are thus not interchangeable in character.

14. *Given r_1 and r_2, the velocities of two rays which are transmitted in the same direction, to find $\lambda\,\mu\,\nu$, the direction-cosines of the line of transmission.*

From Article 8,
$$\frac{\lambda^2}{\frac{1}{r_1^2}-\frac{1}{a^2}} + \frac{\mu^2}{\frac{1}{r_1^2}-\frac{1}{b^2}} + \frac{\nu^2}{\frac{1}{r_1^2}-\frac{1}{c^2}} = 0,$$

$$\frac{\lambda^2}{\frac{1}{r_2^2}-\frac{1}{a^2}} + \frac{\mu^2}{\frac{1}{r_2^2}-\frac{1}{b^2}} + \frac{\nu^2}{\frac{1}{r_2^2}-\frac{1}{c^2}} = 0.$$

Determining the ratios $\lambda^2 : \mu^2 : \nu^2$ from these equations, we find that
$$\frac{\lambda^2}{\left(\frac{1}{b^2}-\frac{1}{c^2}\right)\left(\frac{1}{r_1^2}-\frac{1}{a^2}\right)\left(\frac{1}{r_2^2}-\frac{1}{a^2}\right)}$$
is equal to the two corresponding symmetrical fractions.

Each of the fractions is equal to the fraction of which the numerator is the sum of the three numerators, and the denominator the sum of the three denominators.

The sum of the numerators of the three fractions is unity: the sum of the denominators is

$$\frac{1}{r_1^2 r_2^2}\left[\left(\frac{1}{b^2}-\frac{1}{c^2}\right) + \left(\frac{1}{c^2}-\frac{1}{a^2}\right) + \left(\frac{1}{a^2}-\frac{1}{b^2}\right)\right]$$
$$-\left(\frac{1}{r_1^2}+\frac{1}{r_2^2}\right)\left[\frac{1}{a^2}\left(\frac{1}{b^2}-\frac{1}{c^2}\right) + \frac{1}{b^2}\left(\frac{1}{c^2}-\frac{1}{a^2}\right) + \frac{1}{c^2}\left(\frac{1}{a^2}-\frac{1}{b^2}\right)\right]$$
$$+\frac{1}{a^4}\left(\frac{1}{b^2}-\frac{1}{c^2}\right) + \frac{1}{b^4}\left(\frac{1}{c^2}-\frac{1}{a^2}\right) + \frac{1}{c^4}\left(\frac{1}{a^2}-\frac{1}{b^2}\right).$$

The coefficients of $\dfrac{1}{r_1^2 r_2^2}$ and $\dfrac{1}{r_1^2}+\dfrac{1}{r_2^2}$ vanish, and the remaining term may be transformed into

$$-\left(\frac{1}{a^2}-\frac{1}{b^2}\right)\left(\frac{1}{b^2}-\frac{1}{c^2}\right)\left(\frac{1}{c^2}-\frac{1}{a^2}\right).$$

Hence
$$\lambda^2 = \frac{\left(\frac{1}{r_1^2}-\frac{1}{a^2}\right)\left(\frac{1}{r_2^2}-\frac{1}{a^2}\right)}{\left(\frac{1}{b^2}-\frac{1}{a^2}\right)\left(\frac{1}{c^2}-\frac{1}{a^2}\right)},$$

$$\mu^2 = \frac{\left(\frac{1}{r_1^2}-\frac{1}{b^2}\right)\left(\frac{1}{r_2^2}-\frac{1}{b^2}\right)}{\left(\frac{1}{c^2}-\frac{1}{b^2}\right)\left(\frac{1}{a^2}-\frac{1}{b^2}\right)},$$

$$\nu^2 = \frac{\left(\frac{1}{r_1^2}-\frac{1}{c^2}\right)\left(\frac{1}{r_2^2}-\frac{1}{c^2}\right)}{\left(\frac{1}{a^2}-\frac{1}{c^2}\right)\left(\frac{1}{b^2}-\frac{1}{c^2}\right)}.$$

There are four corresponding directions, namely, $\lambda\mu\nu$, $\lambda\bar{\mu}\nu$, $\lambda\mu\bar{\nu}$, $\lambda\bar{\mu}\bar{\nu}$, symmetrical to the principal planes of the indicatrix.

15. *Given the direction of a line of transmission, to find the velocities of the corresponding rays.*

If RR' (Fig. 8) be parallel to the axis of the cylinder, and R' be the extremity of an axis of the base, it follows from Article **11** that R is a point on the indicatrix corresponding to a ray transmissible along the axis of the cylinder. The corresponding velocity being $\frac{1}{RN} = \frac{1}{R'N'}$, it is seen that the velocities of transmission are inversely proportional to the axes of the base of the cylinder. An analytical solution is given in Art. **8**.

16. The optic bi-radials (*secondary optic axes*).

From Article **15** it follows that if the two velocities of transmission in a given direction are equal, the corresponding tangent cylinder has a circular base. But at *every* point K' (Fig. 8) on the edge of the base of such a cylinder, the normal of the basal section and therefore of the cylinder, and consequently also the normal of the cylinder and therefore of the indicatrix at every corresponding point K of the section conjugate to the ray, intersect the axis of the cylinder perpendicularly, and have the same length intercepted between the surface and the axis: hence *every* point on the conjugate section corresponds to a ray transmissible with the same velocity along the axis of the cylinder: the normals of the indicatrix at these points, and therefore the planes of polarisation, may have any azimuth whatever.

That in the plane AOC there are two directions, and only two, namely those of the lines Os_1, Os_2 (Fig. 7c), for which the two velocities of transmission are equal, has already been proved (Art. 3).

Along each of these lines Os_1, Os_2, rays can thus be transmitted having any azimuth of plane of polarisation whatever, and the velocity of transmission is b for all of them: in the case of calcite and analogous crystals, such properties only belong to that single direction which is termed the optic axis. By reason of this analogy, the directions Os_1, Os_2, have been likewise termed optic axes.

But not being perpendicular to the corresponding ray-fronts, they do not possess all the characters which belong to the optic axis of a uniaxal crystal: from another pair of directions, of which the optical characters are also such as in the case of a uniaxal crystal only belong to the optic axis, they have been distinguished as *Secondary Optic Axes;* and by Sir William Hamilton as *Lines of Single Ray-Velocity*.[1]

In the case of a biaxal crystal, it is experimentally determined that none of the so-called optic axes, primary or secondary, have directions which pass permanently through the same lines of crystalline particles; the lines of particles through which they pass differ with the colour of the light and the temperature of the crystal: hence the so-called optic axes have no material existence, and are in no proper sense of the word *axes of the crystal*.

Where precision of thought and language is necessary, the lines may appropriately be termed *the Optic Bi-radials*, for they are directions in which a line is doubly a radius vector of the ray-surface: the term *uniradial* has already been assigned a distinct signification by Mac Cullagh.[2]

When the indicatrix is a spheroid at all temperatures of the crystal and for all colours of light, the bi-radial is found to be an axis of morphological and physical symmetry, and an axis of revolution of the ray-surface; it always passes through the same line of crystalline particles: such a line may be regarded as a true axis of the crystal.

17. *There cannot be more than one pair of optic bi-radials.*

It has already been proved that Os_1, Os_2, are the only directions for which the velocities of the rays transmissible along the same line, lying in a plane of symmetry of the indicatrix, are equal: it remains to prove that there are no other bi-radials in any direction whatever.

---

[1] *Trans. Roy. Irish Acad.*: 1837, vol. 17, p. 132.
[2] *Ibid.*: 1839, vol. 18, p. 40.

From Article 8 it is seen that the velocity of transmission r is connected with the values of $\lambda\ \mu\ \nu$ by the equation

$$\frac{a^2\lambda^2}{r^2-a^2}+\frac{b^2\mu^2}{r^2-b^2}+\frac{c^2\nu^2}{r^2-c^2}=0;$$

whence

$$a^2\lambda^2(r^2-b^2)(r^2-c^2)+b^2\mu^2(r^2-c^2)(r^2-a^2)+c^2\nu^2(r^2-a^2)(r^2-b^2)=0.$$

Since a, b, c are in descending order of magnitude, the expression on the left-hand side of the last equation is positive, and therefore cannot be zero, if r has any value greater than a or less than c: hence no velocity of transmitted ray can be greater than a or less than c.

Further, if μ is distinct from zero, the above expression is necessarily negative when $r=b$; hence it changes sign and passes through a zero value as r decreases from a to b, and again as r decreases from b to c. If μ is distinct from zero, the two values of r^2 which satisfy the above equation are thus unequal.

Hence the bi-radials can only lie in the plane AOC.

That in the plane AOC there are only two such lines may also be seen from the fact that for any direction lying in this plane one velocity of transmission is always b; when the two velocities are equal the second velocity must also be b: hence if S is a point on the curve $AC\overline{A}\overline{C}$ such that the perpendicular to the radius vector conjugate to OS is equal to OB, the points S correspond to directions Os of single ray-velocity: there are four such points lying at the extremities of two diameters.

The directions Os may also be readily found from the above general equation: for all rays lying in the plane AOC, μ is zero, and the general relation becomes

$$\frac{a^2\lambda^2}{r^2-a^2}+\frac{c^2\nu^2}{r^2-c^2}=0;$$

hence, the rays in this plane for which the two velocities of transmission are both equal to b are given by the equation

$$\frac{a^2\lambda^2}{b^2-a^2}+\frac{c^2\nu^2}{b^2-c^2}=0,$$

which is identical with the equation given in Article 3.

18. *Equation of the planes conjugate to the optic bi-radials.*

The equation of a plane conjugate to a line $\lambda\ \mu\ \nu$ is

$$a^2 x\lambda + b^2 y\mu + c^2 z\nu = 0.$$

For the bi-radials, $\mu=0$ and

$$\frac{a^2\lambda^2}{b^2-a^2}+\frac{c^2\nu^2}{b^2-c^2}=0.$$

Hence their conjugate planes are given by the equation

$$\frac{x^2}{c^2(b^2-c^2)} = \frac{z^2}{a^2(a^2-b^2)}.$$

19. *The direction of a line Or being defined by its inclinations σ_1, σ_2, to the bi-radials Os_1, Os_2, to find the planes of polarisation of the two rays which can be transmitted along it.*

Let $[s_1]$, $[s_2]$, $[r]$ (Fig. 9) be the sections of the indicatrix which are conjugate to the lines Os_1, Os_2, Or respectively, and let D_1, D_2, be two adjacent points of intersection of $[r]$ with $[s_1]$ and $[s_2]$.

D_1 being common to the sections $[r]$ and $[s_1]$, the tangent plane of the indicatrix at D_1 is parallel to both Or and Os_1, and therefore to the plane Ors_1 containing them. Similarly the tangent plane of the indicatrix at D_2 is parallel to the plane Ors_2.

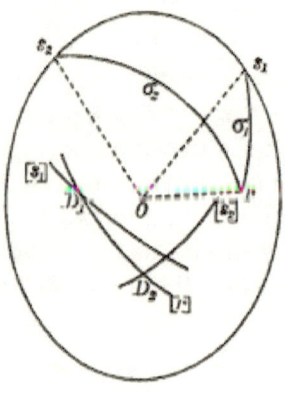

FIG. 9.

Also, all planes tangent to the indicatrix at points on the sections $[s_1]$ and $[s_2]$ are equidistant from the origin (Art. **16**): hence the tangent planes of the indicatrix at D_1 and D_2 are equidistant from the line Or, and are therefore equally inclined to the planes of polarisation, for the latter are the planes of symmetry of the elliptic cylinder which touches the ellipsoid in the section $[r]$ (Art. **12**).

Hence the planes of polarisation of the two rays transmissible in the direction Or are the internal and external bisectors of the angle between the planes Ors_1, Ors_2.

This is the first of the empirical laws of Biot (page 25).

THE OPTIC BI-RADIALS. 47

20. *The direction of a line Or being defined by its inclinations σ_1, σ_2, to the bi-radials Os_1, Os_2, to find the velocities of the two rays which can be transmitted along it.*

If r_1, r_2, be the respective velocities, it follows from the quadratic equation of Art. 8 that

$$\frac{1}{r_1^2}+\frac{1}{r_2^2} = \left(\frac{1}{b^2}+\frac{1}{c^2}\right)\lambda^2 + \left(\frac{1}{c^2}+\frac{1}{a^2}\right)\mu^2 + \left(\frac{1}{a^2}+\frac{1}{b^2}\right)\nu^2$$

$$\frac{1}{r_1^2 r_2^2} = \frac{\lambda^2}{b^2c^2}+\frac{\mu^2}{c^2a^2}+\frac{\nu^2}{a^2b^2}$$

where λ μ ν are the direction-cosines of Or.

It is necessary to express λ μ ν in terms of the angles σ_1, σ_2, and to substitute their values in the above expressions.

Let $l\ 0\ n$, $\bar{l}\ 0\ n$ be the direction-cosines of Os_1, Os_2 respectively, then

$$\cos\sigma_1 = l\lambda+n\nu, \qquad \cos\sigma_2 = -l\lambda+n\nu ;$$

and $\quad 2l\lambda = \cos\sigma_1 - \cos\sigma_2, \qquad 2n\nu = \cos\sigma_1 + \cos\sigma_2.$

Also (from Art. 3)

$$l^2 = \frac{\frac{1}{a^2}-\frac{1}{b^2}}{\frac{1}{a^2}-\frac{1}{c^2}}, \qquad n^2 = \frac{\frac{1}{b^2}-\frac{1}{c^2}}{\frac{1}{a^2}-\frac{1}{c^2}}.$$

Substituting $1 - \lambda^2 - \nu^2$ for μ^2 in the expression for $\frac{1}{r_1^2}+\frac{1}{r_2^2}$, we get

$$\frac{1}{r_1^2}+\frac{1}{r_2^2} = \frac{1}{a^2}+\frac{1}{c^2}-\lambda^2\left(\frac{1}{a^2}-\frac{1}{b^2}\right)+\nu^2\left(\frac{1}{b^2}-\frac{1}{c^2}\right)$$

$$= \frac{1}{a^2}+\frac{1}{c^2}-\left(\frac{1}{a^2}-\frac{1}{c^2}\right)(l^2\lambda^2 - n^2\nu^2)$$

$$= \frac{1}{a^2}+\frac{1}{c^2}+\left(\frac{1}{a^2}-\frac{1}{c^2}\right)\cos\sigma_1\cos\sigma_2.$$

Similarly,

$$\frac{1}{r_1^2 r_2^2} = \frac{1}{a^2c^2}-\frac{\lambda^2}{c^2}\left(\frac{1}{a^2}-\frac{1}{b^2}\right)+\frac{\nu^2}{a^2}\left(\frac{1}{b^2}-\frac{1}{c^2}\right)$$

$$= \frac{1}{a^2c^2}-\left(\frac{1}{a^2}-\frac{1}{c^2}\right)\left(\frac{l^2\lambda^2}{c^2}-\frac{n^2\nu^2}{a^2}\right)$$

$$= \frac{1}{a^2c^2}-\left(\frac{1}{a^2}-\frac{1}{c^2}\right)\left[\frac{(\cos\sigma_1-\cos\sigma_2)^2}{4c^2}-\frac{(\cos\sigma_1+\cos\sigma_2)^2}{4a^2}\right]$$

and

$$\frac{4}{r_1^2 r_2^2} = \frac{4}{a^2c^2}+\left(\frac{1}{a^2}-\frac{1}{c^2}\right)^2(\cos^2\sigma_1+\cos^2\sigma_2)+2\left(\frac{1}{a^4}-\frac{1}{c^4}\right)\cos\sigma_1\cos\sigma_2.$$

Hence
$$\left(\frac{1}{r_1^2}-\frac{1}{r_2^2}\right)^2 = \left(\frac{1}{r_1^2}+\frac{1}{r_2^2}\right)^2 - \frac{4}{r_1^2 r_2^2}$$
$$= \left(\frac{1}{a^2}-\frac{1}{c^2}\right)^2 \sin^2\sigma_1 \sin^2\sigma_2,$$
or $\dfrac{1}{r_1^2}-\dfrac{1}{r_2^2} = \pm\left(\dfrac{1}{a^2}-\dfrac{1}{c^2}\right)\sin\sigma_1\sin\sigma_2.$

If r_1 be the larger velocity, we thus have
$$\frac{1}{r_1^2}-\frac{1}{r_2^2} = \left(\frac{1}{a^2}-\frac{1}{c^2}\right)\sin\sigma_1\sin\sigma_2.$$

This is the second of the empirical laws of Biot (page 25).
From the above we find that
$$\frac{2}{r_1^2} = \frac{1}{a^2}+\frac{1}{c^2}+\left(\frac{1}{a^2}-\frac{1}{c^2}\right)\cos(\sigma_1-\sigma_2)$$
$$\frac{2}{r_2^2} = \frac{1}{a^2}+\frac{1}{c^2}+\left(\frac{1}{a^2}-\frac{1}{c^2}\right)\cos(\sigma_1+\sigma_2):$$
whence
$$\frac{1}{r_1^2} = \frac{1}{a^2}\cos^2\frac{\sigma_1-\sigma_2}{2}+\frac{1}{c^2}\sin^2\frac{\sigma_1-\sigma_2}{2}$$
$$\frac{1}{r_2^2} = \frac{1}{a^2}\cos^2\frac{\sigma_1+\sigma_2}{2}+\frac{1}{c^2}\sin^2\frac{\sigma_1+\sigma_2}{2}.$$

21. *The ray-front corresponding to the ray Or is perpendicular to the transverse plane $RNOr$, and intersects that plane in a line parallel to OR.*

Geometrical Proof.

(a.) As usual, let R (Fig. 10) be the point of the indicatrix corresponding to the ray Or, and RN be the normal of the indicatrix at R: and let $RDEN$ be a plane perpendicular to the plane $RNOr$; it will intersect the indicatrix in an ellipse. Let D be a point on this ellipse distant from R by an arc which is a small quantity of the first order: to this order of small quantities the normal of the ellipse at D is the normal of the indicatrix at that point: if OE be drawn to intersect DE perpendicularly, the line EN is, to the first order of small quantities, perpendicular to the plane $RNOr$ and parallel to DR, and DE is equal to RN. Hence, if Od be the ray corresponding to the point D of the indicatrix, the plane Ord is perpendicular to the plane $RNOr$, and $Od = \dfrac{1}{DE}$, by construction $= \dfrac{1}{RN}$ $= Or$. Hence the line rd is also perpendicular to the plane $RNOr$.

But h and D being adjacent points on the indicatrix, r and d are adjacent points on the ray-surface, and the line rd is thus tangent to the ray-surface at r.

The tangent plane to the ray-surface at r is the plane of the ray-front corresponding to the ray Or: it must pass through all lines tangent to the ray-surface at r, and thus through the line rd, and be perpendicular therefore to the plane $RNOr$.

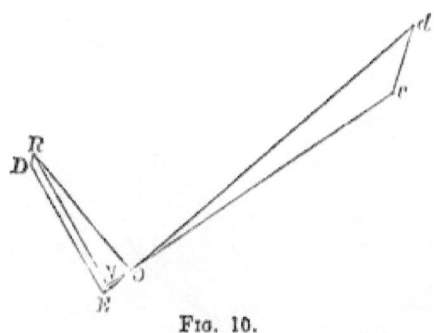

Fig. 10.

(*b.*) The plane $RNOr$ (Fig. 11) also will intersect the indicatrix in an ellipse: let G be a point on this ellipse distant from R by an arc which is a small quantity of the first order: to this order of small quantities GH, the normal of the ellipse at G, is also the normal of the indicatrix at that point. Hence if gOH be perpendicular to GH and $Og = \dfrac{1}{GH}$, Og is the direction of the ray corresponding to the point G, and g is a point on the ray-surface: in the same way as before it follows that rg is a tangent line of the ray-surface, and is thus the intersection of the tangent plane at r with the plane $RNOr$.

Let Or, Og, intersect the ellipse in the points R', G', respectively: then OR' and OG' are respectively conjugate to OR and OG, being perpendicular to RN and GH, and therefore parallel to the tangents at R and G: the area of a parallelogram of which the adjacent sides are conjugate radii vectores is constant; hence
$$RN \cdot OR' = GH \cdot OG'.$$
Also, by construction,
$$RN = \frac{1}{Or}, \quad GH = \frac{1}{Og};$$
hence
$$Or : OR' = Og : OG',$$

and the line rg is therefore parallel to the tangent of the ellipse at R', and consequently to the line OR which is conjugate to OR'.

Hence the ray-front corresponding to the ray Or intersects the transverse plane $RNOr$ perpendicularly in a line parallel to OR.

The diametral line Of, perpendicular to OR and lying in the plane $RNOr$, is therefore normal to the ray-front corresponding to the ray Or (Fig. 12).

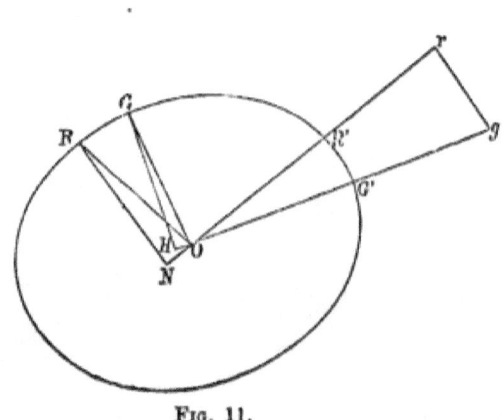

FIG. 11.

Analytical Proof.

The following is interesting to the mathematical student, by reason of the eliminations:—

From Article **7** we have

$$\frac{\frac{x}{r^2-a^2}}{x'} = \frac{\frac{y}{r^2-b^2}}{y'} = \frac{\frac{z}{r^2-c^2}}{z'} = \frac{1}{A} \text{ (say)};$$

hence $\quad x' = \dfrac{Ax}{r^2-a^2},\ y' = \dfrac{Ay}{r^2-b^2},\ z' = \dfrac{Az}{r^2-c^2}.$ (1).

Remembering that

$$\frac{x^2}{r^2-a^2} + \frac{y^2}{r^2-b^2} + \frac{z^2}{r^2-c^2} = 1 \text{ (Art. 7)},$$

we have $\quad xx' + yy' + zz' = A.$ (2).

Also $\quad a^2xx' + b^2yy' + c^2zz' = 0$ (Art. **7**). (3).

It is thus required to determine the tangent plane at a point xyz of the ray-surface in terms of the co-ordinates $x'y'z'$, which are connected by the above equations and also by the relation

$$a^2x'^2 + b^2y'^2 + c^2z'^2 = 1.$$ (4).

Forming the differential of each of the equations (1), we have

$$\left.\begin{array}{l}(r^2-a^2)\delta x' + 2rx'\delta r = A\delta x + x\delta A \\ (r^2-b^2)\delta y' + 2ry'\delta r = A\delta y + y\delta A \\ (r^2-c^2)\delta z' + 2rz'\delta r = A\delta z + z\delta A\end{array}\right\} \quad (5).$$

Multiply these equations by a^2x', b^2y', c^2z', respectively, and add: the quantity δA is thus eliminated, for its coefficient $a^2xx' + b^2yy' + c^2zz'$ vanishes by relation (3); we then have

$$(r^2-a^2)a^2x'\delta x' + (r^2-b^2)b^2y'\delta y' + (r^2-c^2)c^2z'\delta z' + 2r\delta r = A(a^2x'\delta x + b^2y'\delta y + c^2z'\delta z).$$

Remembering that $a^2x'\delta x' + b^2y'\delta y' + c^2z'\delta z' = 0$, owing to relation (4), we have

$$-(a^4x'\delta x' + b^4y'\delta y' + c^4z'\delta z') + 2r\delta r = A(a^2x'\delta x + b^2y'\delta y + c^2z'\delta z).$$

But by Article 4

$$a^4x'^2 + b^4y'^2 + c^4z'^2 = r^2;$$

whence
$$a^4x'\delta x' + b^4y'\delta y' + c^4z'\delta z' = r\delta r.$$

Substituting this value in the preceding equation, we have

$$r\delta r = A(a^2x'\delta x + b^2y'\delta y + c^2z'\delta z)$$

or

$$(x - Aa^2x')\delta x + (y - Ab^2y')\delta y + (z - Ac^2z')\delta z = 0.$$

If $l\ m\ n$ be the direction-cosines of Of, the perpendicular to the tangent plane of the ray-surface at $x\ y\ z$, we must have

$$l\delta x + m\delta y + n\delta z = 0, \text{ whence}$$

$$\frac{l}{x - Aa^2x'} = \frac{m}{y - Ab^2y'} = \frac{n}{z - Ac^2z'}: \quad (6).$$

(*a*.) Each of these quantities is equal to

$$\frac{lx' + my' + nz'}{xx' - Aa^2x'^2 + yy' - Ab^2y'^2 + zz' - Ac^2z'^2}$$

or
$$\frac{lx' + my' + nz'}{xx' + yy' + zz' - A}.$$

The denominator of this expression is zero, by relation (2); hence the numerator $lx' + my' + nz'$ is also zero, for the three equivalent fractions are never all of them indeterminate, and are none of them infinite.

From the relation $\quad lx' + my' + nz' = 0, \quad (7).$
it follows that Of is perpendicular to OR.

(*b*.) Also, multiplying both numerator and denominator of each of the fractions (6) by $y'z'(b^2-c^2)$, $z'x'(c^2-a^2)$, $x'y'(a^2-b^2)$, respectively, we find that each of them is equal to

$$\frac{ly'z'(b^2-c^2) + mz'x'(c^2-a^2) + nx'y'(a^2-b^2)}{y'z'(b^2-c^2)(x-Aa^2x') + z'x'(c^2-a^2)(y-Ab^2y') + x'y'(a^2-b^2)(z-Ac^2z')}.$$

On expanding the denominator, it will be found that the terms involving A mutually destroy each other, owing to the identity
$$a^2(b^2-c^2)+b^2(c^2-a^2)+c^2(a^2-b^2)=0:$$
the denominator thus reduces to
$$xy'z'(b^2-c^2)+yz'x'(c^2-a^2)+zx'y'(a^2-b^2)\,;\text{ or}$$
$$\frac{x'y'z'}{A}\left\{(r^2-a^2)(b^2-c^2)+(r^2-b^2)(c^2-a^2)+(r^2-c^2)(a^2-b^2)\right\}$$
owing to the equations (1).

When multiplied out, this term is likewise found to be zero.

Hence the numerator of the above expression is also zero, and we have the relation
$$ly'z'(b^2-c^2)+mz'x'(c^2-a^2)+nx'y'(a^2-b^2)=0.\quad(8).$$
But this is the condition (Art. 6) that the line Of may lie in the plane RON: hence the front-normal Of lies in the plane $RNOr$ and is perpendicular to OR.

Corollary 1. The inclination of a ray Or to its front-normal Of is the same as the inclination of the normal RN to the radius vector RO at the corresponding point R of the indicatrix (Fig. 12).

Corollary 2. If a ray coincides with the central normal to its ray-front, its direction is perpendicular to an axis of the indicatrix.

Corollary 3. If the ray Or lies in one of the symmetral planes of the indicatrix, the intersection of the corresponding ray-front with the symmetral plane is parallel to the line OR, which is conjugate to the line Or. But if P is any point on an ellipse and Q, $\overline{Q}$, are the extremities of a diameter, the lines PQ, $P\overline{Q}$, are parallel to a pair of conjugate diameters. Hence, if P, Q, $\overline{Q}$, lie in a symmetral plane of the indicatrix, and PQ represents the direction of a ray, the corresponding ray-front is a plane perpendicular to the symmetral plane, and intersects the latter in a line parallel to $P\overline{Q}$.

22. *For the ray Or, the plane of polarisation is perpendicular to the plane containing Or and Of the normal of the corresponding ray-front.*

This follows at once from the last Article, for RN, the normal of the plane of polarisation of the ray, lies in the plane $RNOr$ which has been shown to contain the line Of.

In other words, the plane Orf is the transverse plane for the ray Or.

23. *For the ray Or, corresponding to the point R, the resolved velocity along the normal to the ray-front is measured by the inverse of OR.*

If Of (Fig. 12) be perpendicular to OR and in the plane $RNOr$, and the angle rfO be a right angle, then, by Article **21**, f is the foot of the perpendicular drawn from O to the tangent plane of the ray-surface at r, and Of is the resolved velocity of the ray Or along the normal to the corresponding front.

But Or, Of, are by construction perpendicular respectively to RN and RO: hence the triangles rfO, ONR, are similar, and $\dfrac{RN}{OR} = \dfrac{Of}{Or}$.

Also, by construction, $RN = \dfrac{1}{Or}$; hence $Of = \dfrac{1}{OR}$.

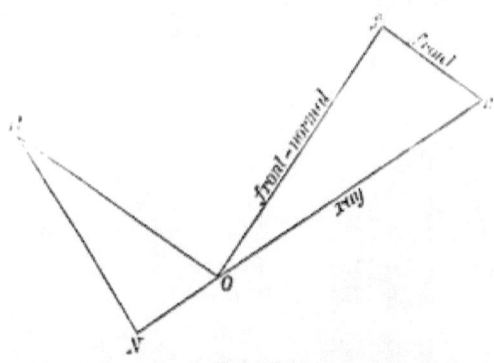

Fig. 12.

24. *The line OR is always a normal of the curve in which the indicatrix is intersected by a central plane parallel to that ray-front which corresponds to the ray Or: in the general case, OR is an axis of the curve.*

RN (Fig. 12) being the normal of the indicatrix at R, any line perpendicular to RN and to the plane $RNOr$ is tangent both to the indicatrix, and to the section of the indicatrix made by any plane which is perpendicular to the plane $RNOr$ at the point R; it is thus tangent to the particular section made by that plane of the series which passes through O. Of this section OR is a central radius vector: hence the tangent at R to the section is at right angles to a central radius vector.

The section being in general an ellipse, R is in such case the extremity of an axis of the section.

Hence it is seen that the ray-surface is the envelope of planes which are distant from a parallel central section of the indicatrix by the inverse lengths of the semi-axes of the latter curve: which is virtually Fresnel's geometrical construction of the surface.

Conversely,

25. *If OR is a central normal (and therefore in general an axis) of the curve in which the indicatrix is intersected by a plane parallel to a given direction of ray-front, the plane through OR normal to the direction of the ray-front contains the ray Or, which corresponds to the point R, and also the line RN, which is the normal of the plane of polarisation of the ray.*

The radius vector OR (Fig. 12) being a central normal of the curve of intersection, a line perpendicular to OR and lying in the plane parallel to the ray-front, is a tangent to the curve of intersection at R: hence RN the normal of the indicatrix at R must lie somewhere or other in the plane ROf perpendicular to this line.

And the ray Or must lie in the same plane.

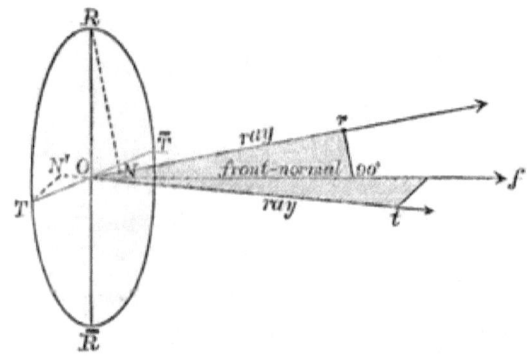

Fig. 13.

26. The two rays corresponding to a given direction of front-normal.

Hence if only the *direction of a ray-front* be given, there are in general two corresponding *positions* of the ray-front, or, in other words, of tangent planes to the ray-surface: and for each there is a corresponding ray (Fig. 13). The rays lie each of them in a plane containing the central normal Of and one of the axes OR, OT, of the section of the indicatrix by a plane parallel to the ray-front; they are thus in two perpendicular planes which intersect in the line Of: the corresponding velocities resolved along the given front-normal are measured by $\frac{1}{OR}$ and $\frac{1}{OT}$ respectively: the normal of the plane of polarisation is parallel to RN for the ray Or, and to TN' for the ray Ot. The directions of vibration at points of the respective rays, according to the latest version of the elastic theory (Art. **1**), are thus indicated by the shading in Figure 13.

It may be remarked that the planes of polarisation of the rays Or, Ot, though perpendicular to the normals RN, TN', are not perpendicular to each other; for it is the lines RO, TO, not the lines RN, TN', which are at right angles: it is easily seen that the cosine of the angle between the planes is equal to $\sin fOr \sin fOt$. Hence only the transverse planes, not the planes of polarisation of the two rays, are perpendicular to each other.

27. The two front-normals corresponding to a given direction of ray.

Similarly, if only the *direction of a ray* be given, there are in general two corresponding *positions and directions* of the ray-front, and two corresponding rays (Fig. 14). The front-normals lie each of them in a plane

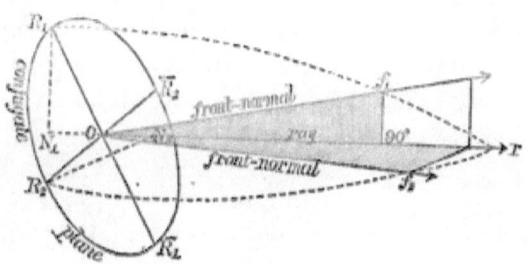

Fig. 14

containing the ray-direction and one of the lines OR_1, OR_2; they are thus in two perpendicular planes which intersect in the line Or: the corresponding ray-velocities are measured by $\dfrac{1}{R_1 N_1}$, $\dfrac{1}{R_2 N_2}$, respectively: the normal of the plane of polarisation is parallel to $R_1 N_1$ for the ray Or_1, and to $R_2 N_2$ for the ray Or_2. The directions of vibration at points of the respective rays, according to the latest version of the elastic theory (Art. **1**), are thus indicated by the shading in Figure 14.

28. *Given the co-ordinates $x'y'z'$ of R, to find θ, the angle between the corresponding ray Or and its front-normal Of.*

Or, Of, being perpendicular to RN, RO, respectively (Fig. 12),

$$\cos\theta = \cos rOf = \cos NRO = \frac{RN}{RO} = \frac{1}{rr'};$$

hence $\tan^2\theta = r^2r'^2 - 1$
$= (a^4x'^2 + b^4y'^2 + c^4z'^2)(x'^2 + y'^2 + z'^2) - (a^2x'^2 + b^2y'^2 + c^2z'^2)^2$
$= (a^2 - b^2)^2 x'^2 y'^2 + (b^2 - c^2)^2 y'^2 z'^2 + (c^2 - a^2)^2 z'^2 x'^2.$

29. *Given the direction-cosines $\lambda\ \mu\ \nu$ of a line of transmission, to find θ, the angle between the corresponding ray Or and its front-normal Of.*

Find r_1^2, r_2^2, (Art. **8**), and then $x_1'y_1'z_1'$, $x_2'y_2'z_2'$, the co-ordinates of the points R_1, R_2 (Art. **9**); also $r_1'^2 = x_1'^2 + y_1'^2 + z_1'^2$ and $r_2'^2 = x_2'^2 + y_2'^2 + z_2'^2$; finally we have $\sec\theta_1 = r_1 r_1'$; $\sec\theta_2 = r_2 r_2'$.

30. *If a ray lies in a given axial plane of the indicatrix, to find the directions for which the inclination θ to the front-normal is a maximum.*

First method.

Let the given axial plane be AOC. Each direction of transmission lying in this plane corresponds to two points R_1, R_2, on the indicatrix: one of these, R_2, always coincides with B, and the corresponding ray coincides with its front-normal; the other, R_1, is in the plane AOC, and the corresponding ray coincides with its front-normal only when R_1 is at A, $\bar{A}$, C, or $\bar{C}$.

If $x'\ 0\ z'$ be the co-ordinates of R_1, $\tan\theta = (c^2 - a^2)z'x'$ (Art. **28**).

Hence, writing $c^2 z'^2 = 1 - a^2 x'^2$, we have

$$\tan^2\theta = \left(\frac{c}{a} - \frac{a}{c}\right)^2 (1 - a^2 x'^2) a^2 x'^2 = \left(\frac{c}{a} - \frac{a}{c}\right)^2 \left\{\tfrac{1}{4} - (a^2 x'^2 - \tfrac{1}{2})^2\right\}.$$

For a maximum value of θ, $a^2 x'^2 = \tfrac{1}{2} = c^2 z'^2$.

Or being parallel to the tangent of the indicatrix at R_1,

$$\tan rOA = -\frac{a^2 x'}{c^2 z'} = \pm \frac{a}{c}$$

or Or is parallel to AC or $\overline{AC}$, when the inclination to the front-normal is a maximum.

Second method.

From Article **21**, Corollary 3, if P is a point on the indicatrix lying in the symmetral plane AOC, and PA represents the direction of a ray belonging to the elliptic section, the corresponding ray-front is perpendicular to the plane AOC and intersects it in a line parallel to $\overline{PA}$; hence the angle $AP\overline{A}$ is the angle of inclination of the ray to its front: the interior angle is a minimum when P coincides with C or $\bar{C}$.

31. *If a ray lies in a given axial plane of the indicatrix, to find the maximum inclination of the ray to its front-normal.*

First method.

Taking AOC for the given axial plane, as in the preceding Article, we have, when $a^2 c'^2 = \frac{1}{2}$,

$$\tan \theta = \pm \tfrac{1}{2}\left(\frac{c}{a} - \frac{a}{c}\right),$$

and $\cot \theta = \tan 2 .ACO$, or $\theta = \frac{\pi}{2} - 2.ACO$.

Hence the maximum or minimum angle which a ray lying in the axial plane AOC can make with its front is given by the angle $AC\bar{A}$.

Second method.

This result is also manifest from the fact that when the ray Or is parallel to $\bar{A}C$, the conjugate diameter OR_1, and therefore also the ray-front, is parallel to CA; as in the second proof given in Art. **30**.

32. *Given the co-ordinates $x'y'z'$ of R, to find the direction-cosines $l\, m\, n$ of the normal Of to the corresponding ray-front.*

For any line $l\, m\, n$ in the plane $RNOr$, as already proved in Article **6**, we have the equation

$$ly'z'(b^2 - c^2) + mz'x'(c^2 - a^2) + nx'y'(a^2 - b^2) = 0.$$

If the line $l\, m\, n$ is likewise perpendicular to OR, of which the direction-cosines are $\frac{x'}{r'}, \frac{y'}{r'}, \frac{z'}{r'}$, we have also

$$lx' + my' + nz' = 0.$$

From these equations the ratios $l : m : n$ are found to be:—

$$\frac{l}{D} = \frac{m}{E} = \frac{n}{F}$$

where

$$D = x'z'^2(c^2 - a^2) - x'y'^2(a^2 - b^2) = x'(1 - a^2 r'^2),$$
$$E = y'x'^2(a^2 - b^2) - y'z'^2(b^2 - c^2) = y'(1 - b^2 r'^2),$$
$$F = z'y'^2(b^2 - c^2) - z'x'^2(c^2 - a^2) = z'(1 - c^2 r'^2);$$

whence

$$\frac{l}{x'(1 - a^2 r'^2)} = \frac{m}{y'(1 - b^2 r'^2)} = \frac{n}{z'(1 - c^2 r'^2)},$$

where $\quad r'^2 = OR^2 = x'^2 + y'^2 + z'^2.$

These equations determine the direction-cosines $l\, m\, n$ of the normal Of of the ray-front corresponding to the point $x'y'z'$ or R.

33. *Given the direction-cosines l m n of the front-normal Of, to find f_1, f_2, the respective velocities of the two corresponding ray-fronts resolved normally to them.*

In Article 23 it was shown that the velocity of the ray-front of the ray Or resolved normally to the front is $\dfrac{1}{Oli}$: denoting the resolved velocity by f and substituting $f = \dfrac{1}{Oli} = \dfrac{1}{r'}$ in the equations of the last Article, we have

$$\frac{l}{\dfrac{f^2-a^2}{x'}} = \frac{m}{\dfrac{f^2-b^2}{y'}} = \frac{n}{\dfrac{f^2-c^2}{z'}}.$$

Each of these fractions is equal to

$$\frac{l\dfrac{l}{f^2-a^2} + m\dfrac{m}{f^2-b^2} + n\dfrac{n}{f^2-c^2}}{lx' + my' + nz'}.$$

But the denominator of the last expression has been proved to be zero (Art. 32); hence the numerator is also zero, for the fractions equivalent to the expression are never all of them indeterminate and are never infinite; we thus have

$$\frac{l^2}{f^2-a^2} + \frac{m^2}{f^2-b^2} + \frac{n^2}{f^2-c^2} = 0.$$

This is a quadratic equation in f^2, and its two roots f_1^2, f_2^2, are the resolved velocities required.

Multiplied out it takes the form
$$f^4 - f^2\{(b^2+c^2)l^2 + (c^2+a^2)m^2 + (a^2+b^2)n^2\} + b^2c^2l^2 + c^2a^2m^2 + a^2b^2n^2 = 0.$$

34. *Given f_1, f_2, the velocities of normal-transmission of two ray-fronts having a common direction of normal Of_1f_2, to find l m n, the direction-cosines of the latter.*

From Art. 33,

$$\frac{l^2}{f_1^2-a^2} + \frac{m^2}{f_1^2-b^2} + \frac{n^2}{f_1^2-c^2} = 0$$

$$\frac{l^2}{f_2^2-a^2} + \frac{m^2}{f_2^2-b^2} + \frac{n^2}{f_2^2-c^2} = 0.$$

Determining the ratios $l^2 : m^2 : n^2$ from these equations, we find that

$$\frac{l^2}{(b^2-c^2)(f_1^2-a^2)(f_2^2-a^2)} = \frac{m^2}{(c^2-a^2)(f_1^2-b^2)(f_2^2-b^2)} = \frac{n^2}{(a^2-b^2)(f_1^2-c^2)(f_2^2-c^2)}.$$

The sum of the numerators of the fractions is unity: the sum of the denominators is

$$f_1^2 f_2^2 [(b^2-c^2)+(c^2-a^2)+(a^2-b^2)]$$
$$-(f_1^2+f_2^2)[a^2(b^2-c^2)+b^2(c^2-a^2)+c^2(a^2-b^2)]$$
$$+a^4(b^2-c^2)+b^4(c^2-a^2)+c^4(a^2-b^2).$$

The coefficients of $f_1^2 f_2^2$ and $f_1^2 + f_2^2$ vanish, and the remaining term may be transformed into

$$-(a^2-b^2)(b^2-c^2)(c^2-a^2).$$

Hence
$$l^2 = \frac{(f_1^2-a^2)(f_2^2-a^2)}{(b^2-a^2)(c^2-a^2)},$$
$$m^2 = \frac{(f_1^2-b^2)(f_2^2-b^2)}{(c^2-b^2)(a^2-b^2)},$$
$$n^2 = \frac{(f_1^2-c^2)(f_2^2-c^2)}{(a^2-c^2)(b^2-c^2)}.$$

There are four corresponding directions, namely, lmn, $\bar{l}mn$, $l\bar{m}n$, $lm\bar{n}$, symmetrical to the principal planes of the indicatrix.

35. *Given the direction-cosines $l\ m\ n$ of the normals Of_1, Of_2, of two ray-fronts having the same direction, to find the co-ordinates $x_1'y_1'z_1'$, $x_2'y_2'z_2'$, of the corresponding points R, T, on the indicatrix.*

The values f_1^2, f_2^2, having been found by the equation of Article 33, the co-ordinates $x_1' y_1' z_1'$, $x_2' y_2' z_2'$ of the points R and T respectively are determined by the following equations, also from Art. 33 :—

$$\frac{x_1'}{\dfrac{l}{f_1^2-a^2}} = \frac{y_1'}{\dfrac{m}{f_1^2-b^2}} = \frac{z_1'}{\dfrac{n}{f_1^2-c^2}};$$

$$\frac{x_2'}{\dfrac{l}{f_2^2-a^2}} = \frac{y_2'}{\dfrac{m}{f_2^2-b^2}} = \frac{z_2'}{\dfrac{n}{f_2^2-c^2}};$$

remembering, also, that the co-ordinates of each point must satisfy the equation of the indicatrix.

It will be observed that the above equations are identical in form with those given in Art. 9: in the one case the direction-cosines and velocities are those belonging to the rays, and in the other case are those belonging to the front-normals.

36. *Given f_1, f_2, the velocities of normal-transmission of two ray-fronts having a common direction of normal $Of_1 f_2$, to find the co-ordinates $x_1'y_1'z_1'$, $x_2'y_2'z_2'$, of the corresponding points R, T, on the indicatrix.*

From Art. 35

$$\frac{x_1'}{\dfrac{l}{f_1^2-a^2}} = \frac{y_1'}{\dfrac{m}{f_1^2-b^2}} = \frac{z_1'}{\dfrac{n}{f_1^2-c^2}} = \frac{1}{A}\ \text{(say)},$$

The square of each of these fractions is equal to
$$\frac{x_1'^2 + y_1'^2 + z_1'^2}{\dfrac{l^2}{(f_1^2-a^2)^2} + \dfrac{m^2}{(f_1^2-b^2)^2} + \dfrac{n^2}{(f_1^2-c^2)^2}}.$$

Hence, remembering that $x_1'^2 + y_1'^2 + z_1'^2 = OR^2 = \dfrac{1}{f_1^2}$, we find on substituting the values of l^2, m^2, n^2, given in Art. **34**,

$$-\frac{A^2}{f_1^2} = \Sigma \frac{f_2^2 - a^2}{(c^2-a^2)(a^2-b^2)(f_1^2-a^2)}$$

or $\dfrac{\Sigma\{(b^2-c^2)(f_1^2-b^2)(f_1^2-c^2)(f_2^2-a^2)\}}{(a^2-b^2)(b^2-c^2)(c^2-a^2)(f_1^2-a^2)(f_1^2-b^2)(f_1^2-c^2)}.$

On expansion of the numerator it will be seen that the terms involving $f_1^4 f_2^2$, $f_1^2 f_2^2$, f_1^4, $a^2 b^2 c^2$, all vanish, and that the coefficients of f_1^2 and f_2^2 are equal but of opposite sign: the numerator then takes the form
$$(f_1^2 - f_2^2)(a^2 - b^2)(b^2 - c^2)(c^2 - a^2)$$

Hence $A^2 = -\dfrac{f_1^2(f_1^2 - f_2^2)}{(f_1^2-a^2)(f_1^2-b^2)(f_1^2-c^2)},$

and $x_1'^2 = \dfrac{l^2}{A^2(f_1^2 - a^2)^2} = -\dfrac{(f_2^2 - a^2)(f_1^2 - b^2)(f_1^2 - c^2)}{f_1^2(f_1^2 - f_2^2)(c^2 - a^2)(a^2 - b^2)}.$

Corresponding expressions give the values of $y_1'^2, z_1'^2, x_2'^2, y_2'^2, z_2'^2$.

The above relation, with many others of this Chapter, was first given by Prof. Sylvester,[1] starting from the vibrational inferences of Fresnel.

37. *Given the direction-cosines l m n of a front-normal Of, to find those of the corresponding rays Or, Ot.*

Find the co-ordinates of R and T by the method of Art. **35**, and then the direction-cosines of the rays Or, Ot, by means of the equations in Article **6**.

38. *Given the direction-cosines of a ray Or, to find those of the corresponding front-normals.*

Find the co-ordinates of R_1 and R_2 (Art. **9**), and then the direction-cosines of the corresponding front-normals by Article **32**.

39. *The front-normal surface, or pedal of the ray-surface.*

It was shown in Art. **33** that if f be the velocity of transmission of a ray-front resolved along its normal Of, of which the direction-cosines are $l\,m\,n$,
$$\frac{l^2}{f^2 - a^2} + \frac{m^2}{f^2 - b^2} + \frac{n^2}{f^2 - c^2} = 0.$$

[1] *Philos. Magazine*, ser. 3: 1837, vol. 11, pp. 461, 537; 1838, vol. 12, pp. 73, 341.

Hence, if $x y z$ be the co-ordinates of f, and the length of Of be denoted by r, we have
$$r = f;\quad x = lr,\ y = mr,\ z = nr.$$
Substituting in the above equation, we get
$$\frac{x^2}{r^2 - a^2} + \frac{y^2}{r^2 - b^2} + \frac{z^2}{r^2 - c^2} = 0.$$

This is the equation of the locus of the points f, or of the pedal of the ray-surface: the velocity of normal-propagation of a ray-front along any radius vector of the surface is measured by the length of the radius vector.

40. *The polar reciprocal of the ray-surface belongs to the same family: surface of wave-slowness or index-surface.*

The radius of a concentric reciprocating sphere being taken as unity, the pole $\xi\,\eta\,\zeta$ which corresponds to the ray-front will lie in the front-normal Of at a distance $\dfrac{1}{Of}$ from the origin.

Hence if $x\,y\,z\,r$ refer to the point f, and $\xi\,\eta\,\zeta\,\rho$ to the pole of the ray-front, we have
$$\frac{x}{r} = \frac{\xi}{\rho},\ \frac{y}{r} = \frac{\eta}{\rho},\ \frac{z}{r} = \frac{\zeta}{\rho};\ r = \frac{1}{\rho}.$$

Substituting these values in the equation of the locus of the points f, we find for the equation of the polar reciprocal of the ray surface
$$\frac{\xi^2}{\frac{1}{\rho^2} - a^2} + \frac{\eta^2}{\frac{1}{\rho^2} - b^2} + \frac{\zeta^2}{\frac{1}{\rho^2} - c^2} = 0$$

or
$$\frac{\frac{1}{a^2}\xi^2}{\rho^2 - \frac{1}{a^2}} + \frac{\frac{1}{b^2}\eta^2}{\rho^2 - \frac{1}{b^2}} + \frac{\frac{1}{c^2}\zeta^2}{\rho^2 - \frac{1}{c^2}} = 0$$

This is a surface of the same family as the ray-surface, being derived from the ellipsoid $\dfrac{x^2}{a^2} + \dfrac{y^2}{b^2} + \dfrac{z^2}{c^2} = 1$ in the same way that the ray-surface itself is derived from the indicatrix $a^2 x^2 + b^2 y^2 + c^2 z^2 = 1$.

The surface has been distinguished by Hamilton as the *surface of wave-slowness*, and by Mac Cullagh as the *index-surface*.[1]

---

[1] *Trans. Roy. Irish Acad.*: 1837, vol. 17, p. 142; 1839, vol. 18, p. 38.

41. The optic bi-normals (*primary optic axes*).

From Article **25** it follows that if the section of the indicatrix by a plane having the direction of the ray-front is a circle, *every* point R on this section corresponds to a ray having the same direction of front and the same resolved velocity normal to the front, and therefore the same position of front. The normals of the indicatrix at the points R, and thus the planes of polarisation of the corresponding rays, may have any azimuth whatever.

That there are at least two directions of central section of the indicatrix for which the curve of section is a circle is seen as follows:—the section of the indicatrix by any plane OBP (Fig. 15), passing through the mean axis OB, is symmetrical both to the line $B O \overline{B}$ and the plane AOC, and therefore to $PO\overline{P}$, P being a point of intersection of the curve with the plane AOC: hence the section is in general an ellipse of which OB, OP,

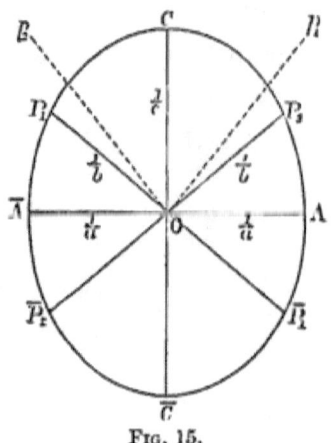

Fig. 15.

are the axes. If the direction OP be so taken in the plane AOC that the radius vector OP is equal to OB, which is always possible since OB is intermediate in length between OA and OC, the **axes** of the ellipse become equal and the ellipse becomes a circle.

There are only two directions for which a radius vector OP of the ellipse AOC has the value OB or $\dfrac{1}{b}$. If $x\,0\,z$ are the co-ordinates of P,

$$a^2 x^2 + c^2 z^2 = 1, \text{ and } x^2 + z^2 = \frac{1}{b^2}: \text{ hence } \frac{x^2}{b^2 - c^2} = \frac{z^2}{a^2 - b^2}.$$

If $l\ 0\ n$ be the direction-cosines of Op, the perpendicular to the line OP,
$$\frac{l}{n} = -\frac{z}{x}, \text{ and } \frac{l^2}{a^2-b^2} = \frac{n^2}{b^2-c^2}.$$

The angle pOC is given by the relation $\tan pOC = \pm \sqrt{\dfrac{a^2-b^2}{b^2-c^2}}$.

The two directions Op_1, Op_2, thus possess certain optical characters which in calcite and analogous crystals only belong to that single direction which is termed the optic axis: for ray-fronts normal to either Op_1 or Op_2 may have any azimuth of plane of polarisation whatever, and their velocities resolved normally to the direction of front are equal. By reason of this analogy, the directions Op_1, Op_2, have likewise been termed optic axes.

But the front-normals Op_1, Op_2, not being coincident with the corresponding rays, for they are not axes of the indicatrix, the directions Op_1, Op_2, do not possess all the characters which belong to the optic axis of a uniaxal crystal; to distinguish them from the directions Os_1, Os_2, which have been termed secondary optic axes, they have received the name *Primary Optic Axes*: they have also been termed by Sir William Hamilton *Lines of Single Normal-Velocity*.

Where precision of thought and language is necessary, the lines may conveniently be termed the *Optic Bi-normals*, for they are directions in which a line is doubly the central normal of a ray-front: the term is correlative to bi-radial, and such a bi-normal cannot be confused with that of a three-dimensional curve.

42. *There cannot be more than one pair of optic bi-normals.*

It has already been proved that the only bi-normals in the plane AOC are Op_1, Op_2: it remains to prove that there are no other bi-normals in any direction whatever.

From Article 33 we have for the relation between l, m, n, f the equation
$$\frac{l^2}{f^2-a^2} + \frac{m^2}{f^2-b^2} + \frac{n^2}{f^2-c^2} = 0,$$
or $\quad l^2(f^2-b^2)(f^2-c^2) + m^2(f^2-c^2)(f^2-a^2) + n^2(f^2-a^2)(f^2-b^2) = 0.$

Since a, b, c are in descending order of magnitude, the expression on the left-hand side of the last equation is positive, and therefore cannot be zero, if f has any value greater than a or less than c; as is otherwise evident from the fact that $f = \dfrac{1}{OR}$, where R is a point on the indicatrix: hence no value of f greater than a or less than c can make the expression zero.

Further, if m is distinct from zero, the above expression is necessarily negative when $f = b$: hence it changes sign and passes through a zero value as f decreases from a to b, and again as f decreases from b to c. If m is distinct from zero, the two values of f^2 which satisfy the above equation are thus unequal.

Hence the bi-normals can only lie in the plane AOC.

Since OB is normal to the plane of polarisation for any ray-front of which the normal lies in the plane AOC, one root of the equation corresponding to such a ray-front is always $f = b$, and this must be the value of the equal roots: the directions of the bi-normals may therefore be found directly from the general equation (Art. 33) as follows:—

For any front-normal in the plane AOC, $m = 0$, and the values of f^2 are given by the equation $\dfrac{l^2}{f^2 - a^2} + \dfrac{n^2}{f^2 - c^2} = 0$: hence, the directions of the front-normals in the plane AOC for which $f = b$ are given by the equation $\dfrac{l^2}{a^2 - b^2} = \dfrac{n^2}{b^2 - c^2}$; which is identical with the equation of last Article.

43. *The direction of a line Of being defined by its inclinations π_1, π_2 to the bi-normals Op_1, Op_2, to find the transverse planes of the two rays of which the corresponding fronts are perpendicular to the given line.*

Let $[p_1]$, $[p_2]$, be the circular sections of the indicatrix perpendicular to the bi-normals Op_1, Op_2 respectively, and let $[f]$ be the central section of the indicatrix parallel to the given direction of ray-front (Fig. 16).

Let $[f]$ intersect $[p_1]$, $[p_2]$, in E_1, E_2, respectively.

All radii vectores in the two circular sections being equal, $OE_1 = OE_2$: and the axes OR, OT, of the elliptical section $[f]$ are therefore the internal and external bisectors of the angle E_1OE_2.

By Art. 26 the two rays Or, Ot, corresponding to the front-normal Of are in the planes fOR, fOT, respectively.

Again, Of is perpendicular to both OE_1 and OE_2;

Op_1, Op_2, are perpendicular to OE_1 and OE_2 respectively.

Hence OE_1 is perpendicular to both Of and Op_1, and therefore to their plane fOp_1;

OE_2 is perpendicular to both Of and Op_2, and therefore to their plane fOp_2.

The planes fOE_1, fOp_1, are thus at right angles: likewise the planes fOE_2, fOp_2.

Hence the planes fOR, fOT, which bisect the angles between fOE_1 and fOE_2, also bisect the angles between fOp_1, fOp_2, the planes which pass through Of and the bi-normals Op_1, Op_2. In other words each of the perpendicular planes, which bisect the angles between the two planes passing through Of and one or other of the bi-normals, contains one of the rays of which the front is normal to Of, and is the *transverse* plane of the contained ray. As already pointed out in Art. 26, it is the transverse planes, not the planes of polarisation, of the two rays corresponding to a single direction of front-normal which are perpendicular to each other.

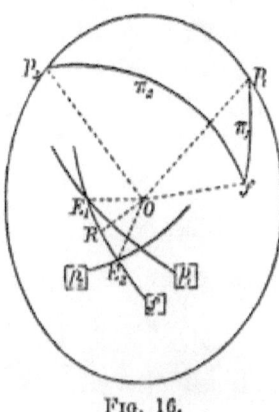

FIG. 16.

44. *The direction of a line Of being defined by its inclinations π_1, π_2, to the bi-normals Op_1, Op_2, to find f_1, f_2, the velocities of normal-transmission of the two ray-fronts which are perpendicular to the given line.*

From the equation of Art. 33, it follows that
$$f_1^2 + f_2^2 = l^2(b^2 + c^2) + m^2(c^2 + a^2) + n^2(a^2 + b^2)$$
$$f_1^2 f_2^2 = l^2 b^2 c^2 + m^2 c^2 a^2 + n^2 a^2 b^2.$$

Hence, proceeding by the method of Art. 20, it may be shown, having due regard to the relative magnitudes of f_1 and f_2, that
$$f_1^2 - f_2^2 = (a^2 - c^2)\sin\pi_1 \sin\pi_2, \text{ and that}$$
$$2f_1^2 = a^2 + c^2 + (a^2 - c^2)\cos(\pi_1 - \pi_2)$$
$$2f_2^2 = a^2 + c^2 + (a^2 - c^2)\cos(\pi_1 + \pi_2):$$
whence
$$f_1^2 = a^2 \cos^2\frac{\pi_1 - \pi_2}{2} + c^2 \sin^2\frac{\pi_1 - \pi_2}{2},$$
$$f_2^2 = a^2 \cos^2\frac{\pi_1 + \pi_2}{2} + c^2 \sin^2\frac{\pi_1 + \pi_2}{2}.$$

45. The bi-radial cone (*the cone of front-normals which correspond to the rays transmissible along a bi-radial*).

We have seen (Article 11) that in general for any given direction of ray Or there are four corresponding points of the indicatrix on two diameters OR_1, OR_2, of the conjugate plane. Also (Article 27), there are two corresponding front-normals Of_1, Of_2, lying in the planes OrR_1, OrR_2, and perpendicular to OR_1, OR_2, respectively. But we have seen that in the case of a bi-radial Os_1, every point S on the conjugate section corresponds to a ray transmissible along that direction with the velocity b. Further, Os_1 is not an axis of the indicatrix, and thus is only coincident with the corresponding central normal On for the two rays transmissible in the direction Os_1 which correspond to the two points B, $\overline{B}$, in which the axis $BO\overline{B}$ meets the conjugate section.

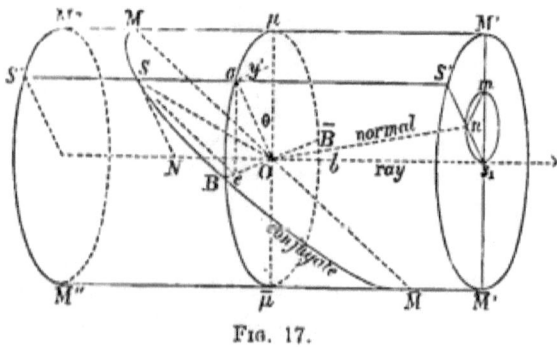

Fig. 17.

If $M''\overline{M}'$ (Fig. 17) be the right circular cylinder touching the indicatrix in the curve $MS\overline{M}$, in which the plane conjugate to the bi-radial Os_1 meets the indicatrix, and $M'S'\overline{M}'$ be a basal section perpendicular to the axis, the front-normal On, corresponding to any point S on the curve, is in the plane $SS's_1$ passing through the axis of the cylinder, and is at right angles to OS. Hence, as the point S moves round the curve $MB\overline{M}$, the corresponding normal On describes a cone of which the bi-radial Os_1 is an edge, for it corresponds to the points B, $\overline{B}$, on the curve: the cone may be conveniently designated a *bi-radial cone*. In the next Article it will be shown that a bi-radial cone intersects the base of its corresponding cylinder in a circle.

Corollary. Since the front corresponding to a ray touches the ray-surface at the point where the ray meets it, a cone of tangent planes can be drawn at each of the points in which a bi-radial meets the ray-surface.

46. *A plane perpendicular to a bi-radial intersects the cone of corresponding front-normals in a circle.*

First proof.

Taking the base of the cylinder at such a distance from O that $Os_1 = b$, we have $ns_1 = Os_1 \tan nOs_1$ (Fig. 17), for the angle $ns_1 O$ is by construction a right angle.

Also, by construction, $SN = \dfrac{1}{Os_1} = \dfrac{1}{b}$.

Hence $\tan nOs_1 = \tan OSN = \dfrac{ON}{NS} = b \cdot ON$,

and $\qquad ns_1 = b^2 \cdot ON$.

Let $B\mu\overline{B}$ be a section of the cylinder parallel to the base, and let the line SS' intersect the curve μB in the point σ: further let MOs_1 be the symmetral plane of the indicatrix which contains the axes OA, OC, and Om be the direction of the front-normal corresponding to the point M on the indicatrix.

SN and σO are equal and parallel, since they are both normal to the axis of the cylinder and in a plane containing it: hence $ON = S\sigma$.

Draw Se, σe, perpendicular to the axis OB; and let the angle between the conjugate section and the base of the cylinder be ϕ: then
$$Se\sigma = MO\mu = mOs_1 = \phi.$$

Let the angle $ns_1 m$ be denoted by θ: to determine the relation between ns_1 and the angle θ, we thus require to express ON or $S\sigma$ in terms of the angle $ns_1 m$ or $\sigma O\mu$.

We have $\quad S\sigma = \sigma e \tan \phi = O\sigma \cos \theta \tan \phi = \dfrac{1}{b} \cos \theta \tan \phi$.

Hence $ns_1 = b^2 \cdot ON = b^2 \cdot S\sigma = b \cos \theta \tan \phi = s_1 m \cos \theta$.

The angle mns_1 is thus a right angle; and the locus of n is a circle passing through the point s_1, and having ms_1 for diameter.

Second proof.

Let $x'y'z'$ be the co-ordinates of any point S on the section conjugate to the bi-radial Os_1: by Article **18**
$$\frac{x'^2}{c^2(b^2-c^2)} = \frac{z'^2}{a^2(a^2-b^2)}.$$

Now y', being the perpendicular from S or σ to the plane $M\mu M's_1$, is equal to
$$\sigma O \sin \mu O\sigma = \frac{\sin \theta}{b}.$$

Finding x' z' also in terms of θ by means of the above equation, we have
$$x'^2 = \frac{b^2-c^2}{a^2-c^2}\frac{\cos^2\theta}{a^2}; \quad z'^2 = \frac{a^2-b^2}{a^2-c^2}\frac{\cos^2\theta}{c^2}.$$

Also $s_1n^2 = On^2 - Os_1^2$.

Draw s_1f parallel to OS and therefore perpendicular to On:

then $On = \dfrac{Os_1^2}{Of} = b^2 \cdot OS$, for $Of = \dfrac{1}{OS}$ (Art. 23).

Hence $s_1n^2 = b^4 OS^2 - b^2$.
$$= b^4 (x'^2 + y'^2 + z'^2) - b^2$$

Substituting the above values of x' y' z' in terms of θ, we get
$$s_1n = \frac{b\cos\theta}{ac}\sqrt{(a^2-b^2)(b^2-c^2)}, \text{ or}$$

$s_1n = k \cos\theta$, where k is constant for all values of θ.

For $\theta = 0$, n takes the position m; hence $s_1n = s_1m \cos\theta$, as before.

Also $s_1m = k = \dfrac{b}{ac}\sqrt{(a^2-b^2)(b^2-c^2)}$.

47. *Aperture of the bi-radial cone.*

If the angle mOs_1 be termed the *aperture* of the cone, the aperture is given by the relation
$$\tan mOs_1 = \frac{s_1m}{s_1O} = \frac{1}{ac}\sqrt{(a^2-b^2)(b^2-c^2)}.$$

The angle mOs_1 is the angle between Os_1, a bi-radial, and Om, a line perpendicular to the plane which is conjugate to that bi-radial; a relation by means of which the above value may likewise be obtained.

48. *Polarisation of the ray corresponding to a given front-normal of the bi-radial cone.*

For the ray transmissible along Os_1 which has On for its front-normal, the normal of the plane of polarisation is SN or ns_1: hence the plane of polarisation of that ray Os_1 which has On for its front-normal, meets the base of the cone in a line parallel to the line nm, or in other words in the line which joins s_1 to the other extremity of that diameter of the circle which passes through n.

49. The bi-normal cone (*the cone of rays corresponding to a front which is perpendicular to a bi-normal*).

In general (Article 26), if OR, OT, (Figs. 13, 18) be the axes of a central section of the indicatrix, the points R, T, correspond to rays Or, Ot, having fronts in the same direction, namely parallel to the plane ORT: also, if Of be the normal to the fronts, the rays Or, Ot, lie in the planes fOR, fOT, and are perpendicular to the lines which are normal to the indicatrix at R and T respectively. But we have seen that all points on the circular section perpendicular to a bi-normal Op_1 correspond to rays having the same position and direction of front, the latter being parallel to the circular section: further Op_1 is not an axis of the indicatrix, and thus is only coincident with the corresponding ray for the two points B, $\bar{B}$, on the circular section.

Hence, as the point R moves round the circular section of the indicatrix, the ray Or, which is always in the plane ROp_1, describes a cone of which the bi-normal Op_1 is an edge, for it corresponds to the points B, $\bar{B}$, on the curve. The cone may be conveniently designated a *bi-normal cone*.

Corollary. Since every front touches the ray-surface where the corresponding ray meets it, a bi-normal is perpendicular to a plane which touches the ray-surface in a closed curve. In the next Article it will be shown that this curve is a circle.

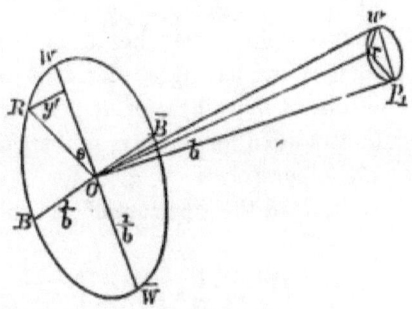

FIG. 18.

50. *A plane perpendicular to a bi-normal intersects the cone of corresponding rays in a circle.*

Let W, $\bar{W}$, (Fig. 18) be the points where the circular section intersects the plane AOC: let $Op_1 = b$, and R be any point on the circular section: the plane ROp_1, containing the ray Or corresponding to the point R, will intersect a plane, drawn through p_1 parallel to the circular section, in a line $p_1 r$ parallel to OR; similarly, if Ow be the ray corresponding to the point W, $p_1 w$ is parallel to OW: hence the angle $rp_1 w =$ angle ROW. Denote it by θ.

Also
$$p_1 r^2 = Or^2 - Op_1^2$$
$$= a^4 x'^2 + b^4 y'^2 + c^4 z'^2 - b^2, \text{ if } x'y'z' \text{ be the co-ordinates of the point } R \text{ (Article 4).}$$

But y' being the perpendicular from R on the plane WOp_1 or AOC, we have
$$y' = OR \sin WOR = \frac{\sin\theta}{b}.$$

Also, since R is on a circular section,
$$\frac{x'^2}{b^2-c^2} = \frac{z'^2}{a^2-b^2} \quad \text{(Article 41)}$$

Finding x' and z' in terms of θ by means of this equation and the relation $a^2x'^2 + b^2y'^2 + c^2z'^2 = 1$, we get
$$x'^2 = \frac{b^2-c^2}{a^2-c^2} \frac{\cos^2\theta}{b^2} \qquad z'^2 = \frac{a^2-b^2}{a^2-c^2} \frac{\cos^2\theta}{b^2}$$

Substituting these values of x'^2, y'^2, z'^2 in the equation for p_1r, we have
$$p_1 r = \frac{\cos\theta}{b} \sqrt{(a^2-b^2)(b^2-c^2)}, \text{ or}$$

$p_1 r = l\cos\theta$, where l is a constant for all values of θ.

For $\theta = 0$, r takes the position w:

hence $p_1 r = p_1 w \cos\theta$; and $p_1 w = \frac{1}{b} \sqrt{(a^2-b^2)(b^2-c^2)}$.

The angle $p_1 rw$ is thus a right angle; hence the locus of r is a circle passing through the point p_1 and having $p_1 w$ for diameter.

The section of the cone of rays by a plane perpendicular to that bi-normal which is the front-normal for the rays is therefore a circle.

51. *Aperture of the bi-normal cone.*

If the angle wOp_1 be termed the aperture of the cone, the aperture is given by the relation
$$\tan wOp_1 = \frac{p_1 w}{p_1 O} = \frac{1}{b^2} \sqrt{(a^2-b^2)(b^2-c^2)}.$$

The angle wOp_1 is equal to the angle between the radius-vector OW and the normal of the indicatrix at W, for Op_1 and Ow are respectively perpendicular to OW and the normal at W; a relation by means of which the above value may likewise be obtained.

52. *Polarisation of the rays of the bi-normal cone.*

The plane of polarisation of the ray Or is a plane perpendicular to the transverse plane $Op_1 r$. The line Op_1, being perpendicular to the plane $p_1 rw$, is perpendicular to the line joining p_1 to r, the other extremity of that diameter of the circle $p_1 rw$ which passes through r; the line rp_1, being perpendicular to both $p_1 O$ and $p_1 r$, is perpendicular to the plane $Op_1 r$ containing them: as any plane passing through $p_1 r$, or its parallel rw, is likewise perpendicular to the plane $Op_1 r$, the plane Orw is the plane of polarisation of the ray Or.

53. Representative surfaces derived from the Indicatrix.

(a) The characters of a ray of light transmitted in a crystal may also be expressed by reference to corresponding points on the polar reciprocal of the indicatrix relative to a concentric sphere: this surface is an ellipsoid represented by the equation $\frac{x^2}{a^2}+\frac{y^2}{b^2}+\frac{z^2}{c^2}=1$ (p. 105 and Fig. 19).

If OR, a radius vector of the indicatrix, be normal to a tangent plane of the polar reciprocal of the indicatrix, meeting the plane in a point M, $OR \cdot OM = 1$, if the radius of the reciprocating sphere be unity. If P be the point in which the tangent plane perpendicular to OR touches the polar reciprocal, and PG be the normal of the latter surface at the point P, the lines PG, PO, lie in the plane $RNOr$: let PG intersect the ray Or in the point G. If m be the point in which OP intersects the plane which touches the indicatrix at R, $OP \cdot Om = 1$, and OP is thus the inverse of RN. Hence, to every point P on the polar reciprocal of the indicatrix corresponds a ray Or: it lies in the plane $PGOr$, and is perpendicular to OP: its velocity of transmission is measured by OP: its transverse plane is $PGOr$: the ray-front intersects the transverse plane $PGOr$ perpendicularly in a line parallel to PG, and its velocity of normal-transmission is measured by PG.

(b) Von Lang[1] has pointed out that if a surface be derived from the ellipsoid $a^2x^2+b^2y^2+c^2z^2=1$ by elongating each radius vector until the new length is measured by the nth power of its original value, the derivative surface may likewise be used for the geometrical representation of the characters of transmitted rays. This result can be generalised still farther, as follows:—

Let $\phi^{-1}(r)$ be any function of r, which always increases and decreases with r, or vice versa: it will have an apsidal (i.e. maximum or minimum) value at the same time as r. If then a new surface be derived by elongating each radius vector r of the indicatrix to a length ρ, determined by the relation $\rho = \phi^{-1}(r)$ or $r = \phi(\rho)$, a central section of the new surface will have its apsidal diameters in exactly the same directions as those of the section of the indicatrix by the same plane. If ρ_1, ρ_2, be the half-lengths of the new diameters, the corresponding ray-fronts are respectively at distances $\frac{1}{r_1}$ and $\frac{1}{r_2}$ or $\frac{1}{\phi(\rho_1)}$ and $\frac{1}{\phi(\rho_2)}$ from the central section; the ray-surface itself is the envelope of these planes.

---

[1] Sitz. Ak. Wien, 1861, vol. 43, sec. 2, p. 645.

The general equation of the new surface is easily found:—

If r be the length of a radius vector of the indicatrix and $l\ m\ n$ be its direction-cosines, $a^2 l^2 + b^2 m^2 + c^2 n^2 = \dfrac{1}{r^2}$; $\xi\ \eta\ \zeta$ being the co-ordinates of the corresponding point on the new surface, $\xi = l\rho,\ \eta = m\rho,\ \zeta = n\rho$:

whence $a^2\xi^2 + b^2\eta^2 + c^2\zeta^2 = \dfrac{\rho^2}{r^2}$

or $a^2\xi^2 + b^2\eta^2 + c^2\zeta^2 = \dfrac{\rho^2}{[\phi(\rho)]^2}$:

which is the required equation.

Fresnel's "surface of elasticity" is the particular case in which $\phi(\rho) = \dfrac{1}{\rho}$, for the equation then becomes $a^2\xi^2 + b^2\eta^2 + c^2\zeta^2 = (\xi^2 + \eta^2 + \zeta^2)^2$.

For the "surface of elasticity," the transverse planes of the rays corresponding to a given direction of ray-front pass through the apsidal diameters of a central section, as in the case of the indicatrix, but the distance of the ray-front corresponding to a semi-diameter of length ρ_1 is not $\dfrac{1}{\rho_1}$ as in the indicatrix, but $\dfrac{1}{\phi(\rho_1)}$ or ρ_1.

The corresponding ray is only perpendicular to the corresponding normal of the representative surface in the case of the indicatrix: in every case, however, the normal of the representative surface lies in the plane passing through the corresponding diameter and the front-normal: for the curves of intersection of the two surfaces by the given plane have parallel tangents at the extremities of their maximum and minimum diameters. Hence, as in the case of the indicatrix, the plane passing through a diameter and a normal of the surface at the extremity of the diameter is the *transverse* plane of the corresponding ray.

(*c*) In exactly the same way a series of surfaces can be derived from the polar reciprocal of the indicatrix.

The above generalisation serves as a reminder that there is not *necessarily* a simple relation between a surface of geometrical representation and the characters of the ether.

CHAPTER V.

Various Optical Relations which are independent of the Physical Character of the Periodic Change.

1. In Chapter II we have shown that after the discovery of the polarisation of light by reflection by Malus in 1808, and of the correspondence of optical and morphological symmetry by Brewster in 1819, the true laws of transmission of light in biaxal crystals must soon have been suggested, independently of any hypothesis as to the physical character of the periodic change: in fact, their enunciation by Fresnel in 1821 was only two years later than Biot's discovery of two empirical laws by which the accuracy of a geometrical representation could be tested. If the truth of the construction given by Huygens for the case of calcite is acknowledged, the suggestion presents itself as soon as the planes of polarisation of the two rays transmissible in any direction in a crystal of calcite are represented by their normals.

In the present Chapter we proceed to indicate very briefly, for the convenience of the student, various other important relations, which, though really independent of any hypothesis as to the nature of the periodic change, are usually imagined and expressed as belonging to an elastic ether. It will at the same time be shown that the form of the ray-surface for biaxal crystals is not merely suggested by a geometrical generalisation as a tentative one, but is a necessary consequence of the difference of symmetrical development of the same physical characters, whatever they may be, which originate the sphere and spheroid of a uniaxal crystal: it will further be shown that the same form of the ray-surface would result from the general features of perpendicularly transverse vibrations, and be independent of the real nature of the periodic change.

Preliminary algebraical expression for the transmission of a ray of common light.

2. It will be convenient, in the first place, to find a mathematical expression connecting the magnitude of the disturbance or change of state at any point in a ray of common light of simple colour with the position of the point, the time, and the period of the vibration. For this purpose it is necessary to make an assumption as to the law of the change: the simplest which can be made is that, at any point of a ray of common light

of simple colour, the variation of the state with the time follows the same law as the variation of position of an isochronous pendulum.

It will be found that, for a ray of simple colour, the expression

$$y = a\sin\left\{\frac{2\pi}{\lambda}(vt-x)+\alpha\right\}$$

is one which satisfies this condition and is consistent with all experiments *as yet referred to*;

x denoting the distance of any point in the ray from a fixed point in it,

y the magnitude of the disturbance or change of state at the point x at the time t;

v the velocity of transmission,

λ the wave-length,

a and α two constants for all values of x and t:—

1. At a given *point*, indicated by its distance x from the origin, the change of state varies periodically with the time t: the same value of y, and therefore the same change of state, recurs whenever the expression $\frac{2\pi}{\lambda}(vt-x)+\alpha$ increases by 2π, that is when t increases by the constant interval $\frac{\lambda}{v}$. The same change of state recurs, but with opposite sign, whenever the expression $\frac{2\pi}{\lambda}(vt-x)+\alpha$ increases by π, that is when t increases by half the above interval.

2. At a given *instant*, indicated by the time t, the change of state is the same in magnitude and sign for all points separated from each other by the distance λ: it is the same in magnitude and opposite in sign for all points separated from each other by half that distance.

3. The relation between y and t is identical with the relation between the position of an isochronous pendulum and the time.

a, being the maximum value of y, is the amplitude of the vibration.

$\frac{2\pi}{\lambda}(vt-x)+\alpha$ being the phase of the vibration at the point x at the time t, α is the phase of the vibration at the origin ($x=0$) at the epoch from which the time is measured ($t=0$).

The period of the vibration being independent of the amplitude, the law is consistent with the independence of colour and intensity.

Conversely, if the magnitude of the change of state at each point of a line is given by the expression $y = a\sin\left\{\frac{2\pi}{\lambda}(vt-x)+\alpha\right\}$, and the change is of the physical character which belongs to light, a ray of light of simple

PRELIMINARY REPRESENTATION OF THE PERIODIC CHANGE.   75

colour is passing along the line with a velocity v: the intensity corresponds to the amplitude a, the colour to the period $\frac{\lambda}{v}$, while the phase of the vibration at the origin at the initial epoch is a.

From analogy with sound, we may tentatively assume that the intensity of the light corresponding to this simple change of state is measured by the square of the amplitude.

Resultant effect of the simultaneous transmission of two or more such rays along the same line.

3. The fact of the periodicity of the change was deduced from experiments relative to the mutual interference of rays of light: it is easily seen that the above expression for the change, combined with the principle of superposition, is consistent with the observed phenomena from which it was deduced.

(a) *If the component rays have the same wave-length and velocity.*

1. For let two rays of the same simple colour, transmissible along a given line with the same velocity v, be represented respectively by the expressions

$$y = a \sin \left\{ \frac{2\pi}{\lambda} (vt - x) + a \right\}$$

$$y = b \sin \left\{ \frac{2\pi}{\lambda} (vt - x) + \beta \right\}:$$

if both are transmitted simultaneously, the principle of superposition requires the resultant change to be determined by the expression

$$y = a \sin \left\{ \frac{2\pi}{\lambda} (vt - x) + a \right\} + b \sin \left\{ \frac{2\pi}{\lambda} (vt - x) + \beta \right\}.$$

If the terms can be added together *in the same way as numerical quantities of a single kind*,

$$y = (a \cos a + b \cos \beta) \sin \frac{2\pi}{\lambda} (vt - x) + (a \sin a + b \sin \beta) \cos \frac{2\pi}{\lambda} (vt - x)$$

$$= c \sin \left\{ \frac{2\pi}{\lambda} (vt - x) + \gamma \right\},$$

if $c^2 = a^2 + b^2 + 2ab \cos(a - \beta)$

and $\tan \gamma = \dfrac{a \sin a + b \sin \beta}{a \cos a + b \cos \beta}.$

Hence the resultant effect of the two rays is identical with that of a single ray transmitted along the same line with the same velocity and the same wave-length (and thus of the same colour), but having an intensity c^2 and an original phase γ. And the intensity of the resultant ray depends not only on the intensities of the component rays but on their difference of phase at the same point at the same instant: if a and b are equal and a differs from β by any odd multiple of π, the intensity of the single resultant ray is constantly zero.

2. It may in this way be shown that the resultant effect of the simultaneous transmission of any number of such rays of the same simple colour along the same line with the same velocity is identical with that of a single ray of the same colour and velocity, and having a determinable phase and intensity.

(b) *If the component rays differ in wave-length or velocity.*

On the other hand, if the component rays differ either in velocity or wave-length, the resultant effect is not that of a single ray of simple colour: the resultant effect is still expressed by

$$a \sin\left\{\frac{2\pi}{\lambda}(vt-x)+a\right\} + b \sin\left\{\frac{2\pi}{\lambda'}(v't-x)+\beta\right\};$$

but the expression cannot take the simpler form $c\sin\left\{\frac{2\pi}{\lambda''}(v''t-x)+\gamma\right\}$, in which c and γ are both constants: indeed, the resultant effect is not periodic at all unless the ratio $\frac{v}{\lambda} : \frac{v'}{\lambda'}$ is commensurable.

Kinematical representation of the periodic change at any point of such a ray.

4. Whatever be the physical character of the periodic change at any point of a ray of light, the state at any point P at a given instant may thus (consistently with any facts as yet indicated) be represented by the above expression

$$y = a \sin\left\{\frac{2\pi}{\lambda}(vt-x)+a\right\}:$$

this algebraical expression may in turn be represented geometrically; the magnitude y being represented by the distance of a point p from the point P, and the distance being considered positive or negative according to the direction in which it is measured. The phenomena of interference, from which the above expression has been deduced, merely require the direction in which the line Pp is measured to be necessarily the same for all points of the same ray, and for all interfering rays transmitted along the same line.

This mode of representation in no way assumes that the *actual* change of state at the point P is a to-and-fro motion of a particle of ether in the arbitrary line Pp; the direction of the line Pp is required to be constant merely to secure that the changes, if they have any directional character at all, may be added together like simple numerical quantities of the same kind: that the change is really directional in character may be inferred from the fact that it is being transmitted in a definite direction through the medium. In exactly the same way, the transmission of a ray of light along a line is sometimes conveniently represented (in the discussion of aberration, for instance) by the transmission of a point along the line with constant velocity, although light is certainly not due to the transmission of a particle along the direction of the ray.

Preliminary algebraical expression for the transmission of a ray of plane-polarised light.

5. It was found by Fresnel, in conjunction with Arago, that two rays of plane-polarised light, if their planes of polarisation are parallel, *may* mutually interfere in exactly the same way as ordinary light: hence, as far as this experiment goes, the periodic change at any point of a plane-polarised ray can be represented in exactly the same way as for common light; the only difference being that while a common ray is so far analogous to a circular cylinder that its characters are identical on all its sides, a plane-polarised ray is analogous to an elliptical cylinder to the extent that the properties of the ray are dissimilarly symmetrical relative to two perpendicular planes (pages 12 and 17).

If all the characters of a plane-polarised ray can be accounted for by such a kinematical representation as is mentioned above, the line Pp must lie either in the plane of polarisation or the transverse plane ; but it may have any inclination whatsoever to the ray, so long as for two interfering rays the direction is identical.

More general representation of the periodic change at any point of a common or plane-polarised ray.

6. Since, as far as the above experiments are concerned, the inclination of the direction Pp to the line of transmission of either a plane-polarised or a common ray, may be any whatsoever, it follows that the change may really not be simple, but *multiple* in direction; assuming that each transmitted periodic change will interfere for itself, as if those having other directions did not exist.

In fact, it will be seen that the periodic change may likewise be represented by the composite expression

$$y = a_1 \sin\left\{\frac{2\pi}{\lambda}(vt-x) + a_1\right\} + \ldots\ldots + a_n \sin\left\{\frac{2\pi}{\lambda}(vt-x) + a_n\right\}$$

consisting of any number of terms: for each separate term, independently of its directional relations, resumes its original value at distances along the ray separated from each other by the common length λ, or at times separated from each other by the common period $\dfrac{\lambda}{v}$: hence, if two rays annihilate each other under given circumstances, annihilation will again take place if one of the rays is moved *parallel to itself* through the distance λ along its line of transmission.

And it is important to remark that as each term recurs *individually* after the same interval of time or distance, the whole expression likewise recurs and has the same total value, even if the terms are *not* subject to the same law of addition as simple numerical quantities.

It will also be obvious on reflection that any ray which is within the reach of experiment is necessarily composite as regards the origin of its vibration, even if it be simple as regards its colour: the luminous source is not a geometrical point, but a surface of considerable dimensions as compared with the wave-length of a ray of light; hence the periodic change, of which the effects are observed at a given point of a line of transmission, is really of composite origin and due to the superposition of the periodic changes transmitted from the points of a luminous area of appreciable magnitude.

As for the difference between a common and a plane-polarised ray, the first suggestion which presents itself is that the latter is due to the distortion of the common ray from which it was derived; just as an elliptical cylinder may be derived from a circular cylinder by compression in a direction inclined to the axis.

Experimental discovery made by Fresnel and Arago.

7. (*a*) But Fresnel and Arago found that, when one of two interfering plane-polarised rays is turned through a right angle round its direction of transmission, the interference-effects completely disappear, whatever the difference of phase of the two rays. Hence, with this relative position of the planes of polarisation, the periodic change produced at any point by the transmission of one ray is in no direction coincident with a periodic change produced by the transmission of the other ray; for as we have seen (Art. 3), such coincidence would involve a variation of intensity of the resultant effect: if this be granted, it follows that for a plane-polarised ray the actual periodic change must be in only a single direction, and the single direction must be perpendicular to the line of transmission; for otherwise the two positions of the plane-polarised ray would give two positions of the periodic change which would have a resolved part in

common. Since the direction is single, it must be in one of the symmetral planes of the ray: hence the direction of the actual periodic change is perpendicular to the direction of transmission, and may be either in or perpendicular to the plane of polarisation: in either case it may be *represented* by a line perpendicular to the plane of polarisation.

In the above experiments of Fresnel and Arago, the rays were allowed to interfere during aerial transmission; it may reasonably be assumed, however, that the same kind of symmetry with respect to two perpendicular planes obtains for a plane-polarised ray as transmitted within any crystalline medium: the assumption is not only reasonable on general grounds, but is consistent at once with all known experimental results and with the requirements of the most recent version of the elastic theory (see also pages 17, 32). It is not the only assumption which can be made: Fresnel himself was led by the hypothesis of an incompressible elastic ether to infer that a plane-polarised ray transmitted within a bi-refractive medium is in general symmetrical to only a single plane, perpendicular to the plane of polarisation; he inferred, in fact, that the vibrations of the ether lie in the transverse plane and are in general oblique, not perpendicular, to the direction of the ray. That Fresnel felt the unsatisfactory character of the inference, in the absence of any experimental proof of the obliquity, will be seen on reference to the original memoir.[1]

(*b*) If the two polarised rays which have been obtained from a ray of common light by means of a crystal of calcite are transmitted along the same line, it is found that the resultant effect is again that of a single ray of common light: hence we may infer that in common light, as in plane-polarised light, the vibrations are perpendicular to the direction of transmission of the ray.

Representation of the resultant effect of the simultaneous transmission along the same line of two or more plane-polarised rays having different directions of planes of polarisation.

8. (*a*.) *If the component rays have the same wave-length and velocity.*
(1.) The periodic change at any point of a plane-polarised ray being kinematically represented by a vibration perpendicular to the plane of polarisation, let two rays be transmitted with the same velocity along the same line, having different directions of the plane of polarisation: and, in the first place, let the algebraical expressions for the corresponding changes be respectively

$$y = a \sin\left\{\frac{2\pi}{\lambda}(vt - x) + a\right\} \text{ and } z = b \sin\left\{\frac{2\pi}{\lambda}(vt - x) + \beta\right\}.$$

---

[1] *Loc. cit.*; 1827, p. 158.

Assuming as before the principle of superposition, the effect of transmitting both rays simultaneously will be represented by the motion of a point of which the co-ordinates y and z, measured along the normals of the planes of polarisation, are given by the expressions

$$y = a \sin \left\{ \frac{2\pi}{\lambda}(vt - x) + a \right\}$$

$$z = b \sin \left\{ \frac{2\pi}{\lambda}(vt - x) + \beta \right\}.$$

Eliminating $\frac{2\pi}{\lambda}(vt - x)$, we find

$$\frac{y^2}{a^2} + \frac{z^2}{b^2} - \frac{2yz}{ab} \cos(a - \beta) = \sin^2(a - \beta).$$

Hence the point, of which the position at any instant represents the resultant disturbance at that instant at a corresponding point on the line of transmission, describes in general an ellipse, of which the magnitude and position relative to the planes of polarisation of the original rays are independent both of x and t: all the ellipses are thus equal and parallel, and form a cylinder of which the base is elliptical, and the axis is in the direction of transmission. It will be found that the direction in which the point moves round the ellipse is determined by the relative phases of the two rays. The composite or resultant ray of light due to the co-existence of the original rays is said to be *elliptically polarised*; a ray of which the characters are related to a cylinder with elliptical base must differ from a ray of common light, of which the characters are the same on all its sides.

(2.) If the rays have the same intensity, and their difference of phase is measured by the angle between their planes of polarisation, $a = b$, and $a - \beta$ is equal to the angle between the directions of y and z: in this case the ellipse becomes a circle, and the cylinder becomes one with a circular base. The composite ray is then said to be *circularly polarised*. Such a ray is similarly related to every plane passing through it, and yet differs from one of common light: for the motion of the representative point is not symmetrical to a plane, and the characters of the ray may conceivably differ with the direction in which the circle is described by the ideal point. In fact, experimental methods enable us to distinguish, not only between a ray of common light and one which is circularly polarised, but between two circularly polarised rays of which the motion of the ideal point is in opposite directions.

(3.) If $\sin(a - \beta) = 0$, that is to say, if the difference of phase is zero or a multiple of π, the ellipse becomes one or other of the two straight lines

$\left(\dfrac{y}{a} \pm \dfrac{z}{b}\right)^2 = 0$: hence the resultant ray is itself *plane-polarised*; the direction of the plane of polarisation depending on the ratio $a:b$, and thus being determined by the relative intensities of the two component rays. Conversely, such a single plane-polarised ray of simple colour is equivalent in its effects to two such plane-polarised rays of the same simple colour, transmitted along the same line with the same velocity, and with their planes of polarisation in *any* assigned directions. If the two assigned directions be perpendicular to each other, and θ be the inclination of one of them to the plane of polarisation of the original ray supposed to be represented by the expression

$$y = a \sin\left\{\frac{2\pi}{\lambda}(vt - x) + \alpha\right\},$$

the two equivalent rays are represented respectively by the expressions

$$y = a \sin\theta \sin\left\{\frac{2\pi}{\lambda}(vt - x) + \alpha\right\}$$

$$z = a \cos\theta \sin\left\{\frac{2\pi}{\lambda}(vt - x) + \alpha\right\};$$

for the resultant effect of these two rays is such that

$$\frac{y}{z} = \tan\theta, \text{ a constant quantity},$$

whatever be the time or the position of the point in the line of transmission.

(4.) Further, it will be seen that any number of such rays of the same simple colour transmitted along the same line with the same velocity but with different phases, amplitudes and planes of polarisation, will have a resultant effect identical in general with that of a single elliptically polarised ray of the same simple colour, transmitted along the same line with the same velocity. For let the simple rays be severally represented by the expressions

$$y_1 = a_1 \sin\left\{\frac{2\pi}{\lambda}(vt - x) + \alpha_1\right\}$$

$$y_2 = a_2 \sin\left\{\frac{2\pi}{\lambda}(vt - x) + \alpha_2\right\}$$

$$\dots\dots\dots\dots\dots\dots\dots\dots$$

$$y_n = a_n \sin\left\{\frac{2\pi}{\lambda}(vt - x) + \alpha_n\right\},$$

and let the inclinations of the respective planes of polarisation to a fixed plane of reference through the line of transmission be $\theta_1, \theta_2, \ldots \theta_n$. Each single ray being equivalent in effect to two rays with perpendicular planes of polarisation, one of them coincident with the fixed plane of reference, the whole system of rays is equivalent to the following two systems:—

$$y = a_1 \sin\theta_1 \sin\left\{\frac{2\pi}{\lambda}(vt-x)+a_1\right\} + \ldots\ldots + a_n \sin\theta_n \sin\left\{\frac{2\pi}{\lambda}(vt-x)+a_n\right\}$$

$$z = a_1 \cos\theta_1 \sin\left\{\frac{2\pi}{\lambda}(vt-x)+a_1\right\} + \ldots\ldots + a_n \cos\theta_n \sin\left\{\frac{2\pi}{\lambda}(vt-x)+a_n\right\};$$

all the members of each of these systems having a common direction of plane of polarisation.

As each system is equivalent to a single plane-polarised ray (Arts. 3 and 5), the two systems are together equivalent in general to a single elliptically polarised ray.

(5.) Whether the resultant ray be elliptically, circularly, or plane-polarised, the resultant change has the same period as the change for each component ray, and is thus of unaltered colour.

(6.) At a given instant, the ideal points representing the state at all points of the resultant ray lie on a spiral curve surrounding the elliptical or circular cylinder, if the ray be elliptically or circularly polarised, and on an undulating curve (the curve of sines) in the transverse plane, if the ray be plane-polarised.

(*b.*) *If the component rays differ in wave-length or velocity.*

If the two component rays differ in wave-length or velocity of transmission, the resultant effect is still represented by the combined expressions

$$y = a \sin\left\{\frac{2\pi}{\lambda}(vt-x)+a\right\}$$

$$z = b \sin\left\{\frac{2\pi}{\lambda'}(v't-x)+\beta\right\}:$$

but it is not periodic at all unless the ratio $\frac{v}{\lambda} : \frac{v'}{\lambda'}$ is commensurable: and even in that case the curve described by an ideal point is not a conic section.

The resultant effect can only be that of a plane-polarised ray if the ratio of y to z, and therefore of $\sin\left\{\frac{2\pi}{\lambda}(vt-x)+a\right\}$ to $\sin\left\{\frac{2\pi}{\lambda}(vt-x)+\beta\right\}$ is independent of the time: but if either v or λ is different from v' or λ' respectively, this constancy is impossible, whether the planes of polarisation of the original rays are real or imaginary.

Discrepancy of observed and calculated results.

9. But the above calculation of the resultant effect of the simultaneous transmission along the same line of two plane-polarised rays of the same colour with planes of polarisation at right angles to each other is in direct disagreement with the experimental result recorded in Art. 7b, for the result of superposition of the two plane-polarised rays obtained from a ray of common light by means of a bi-refractive crystal is not an elliptically polarised ray, but a ray of common light having identical characters on every side. We are thus led to inquire how far the constancy of character of the periodic changes at points in the same ray has really been established by experiment.

In fact, the annihilation-effect (p. 10) of two rays of identical character has only been established for a transference of one of the rays through a distance of at most 50,000 wave-lengths: the wave-length in air for sodium-light being nearly $\frac{3}{5000}$ millimetres, the above distance is nearly 30 millimetres or about one inch: as light is transmitted through air at the rate of 186,000 miles a second, a distance of one inch corresponds to the lapse of only $\frac{1}{11,784,960,000}$ th part of a second.

The discrepancy disappears if a ray is assumed to consist of a series of independent sets of waves of the same length.

10. For the sake of a numerical example, let us imagine that two given rays are absolutely identical in character; that each ray consists of a series of sections; that each section consists of at least a million similar waves, but that the waves of one section are absolutely independent of those of every other, except that they have the same period and are transmitted with the same velocity.

Let the constant sections of one ray be A_1B_1, B_1C_1, C_1D_1........Y_1Z_1, and the identical sections of the other ray be A_2B_2, B_2C_2, C_2D_2........Y_2Z_2: consider the resultant effect of transmitting both heterogeneous rays simultaneously along the same line.

(1.) If the initial points A_1, A_2 coincide, the vibrations are in unison at every point of every section, notwithstanding the heterogeneity of each ray.

(2.) If the ray A_2.........Z_2 be moved parallel to itself along its own direction through the distance $\frac{\lambda}{2}$, the two rays will annihilate each other at all points where identical sections are superposed, but will in general fail to do so in the regions where different sections overlap; that is, for a distance $\frac{\lambda}{2}$ at the end of every section. Hence at any given point there will be annihilation while at least 999,999½ waves pass by, and more or less unison while half a wave is passing the same point.

(3.) In the same way, if the ray $A_2\ldots\ldots\ldots Z_2$ be moved parallel to itself along its own direction through the distance 50,000½ wave-lengths, the two rays will still annihilate each other at all points where identical sections are superposed, but will in general fail to do so in the regions where different sections overlap; that is, for a distance 50,000½ wave-lengths at the end of each section. Hence, at any given point, there will be complete annihilation while at least 949,999½ waves pass by, and more or less unison while 50,000½ waves are passing the same point: in other words, instead of complete annihilation, there is more or less light during at most $\frac{1}{20}$th part of the time: the light will be apparently continuous, but its intensity will not exceed the $\frac{1}{20}$th part of the maximum joint effect of the two rays. The variability of the periodic character will thus account for the appreciable diminution of the interference-effect when one of the rays is moved parallel to itself through a considerable number of wave-lengths.

In the following pages we shall only need to consider sets of waves belonging to a single section of constant periodic character, and may thus proceed as if the constancy of character were really a property of the whole ray.

The same assumption accounts for the remarkable fact that rays of the same simple colour, but obtained from different sources, cannot be made to annihilate each other.

11. Hitherto, for simplicity, we have left unmentioned the remarkable fact that rays of light of the same simple colour, whether common or plane-polarised, cannot be made to annihilate each other if they have been derived from different sources. This is quite inexplicable if a ray is assumed to have constancy of periodic character throughout its extent; but it is immediately accounted for by the assumption arrived at in the preceding Article: if a ray consists of a series of *independent* sets of waves, it is physically impossible for two rays from different sources to be identical in their characters.

For a plane-polarised ray, only the amplitudes and phases will differ in the different sets.

We have seen that two plane-polarised rays of constant periodic character throughout would give an elliptically polarised ray of which the ellipses would have a definite magnitude and position dependent on the amplitudes and phases of the component rays: if each of the plane-polarised rays, instead of being of constant periodic character throughout, consists of independent sets of waves, the resultant effect will generally be a rapid succession of elliptically polarised sets, the magnitudes and positions of the ellipses changing as different sections of the plane-polarised rays become superposed; the resultant ray will thus be generally identical in character on all its sides, as far as observation can detect.

Not only is the assumption of variability of periodic character necessary, but a constancy of periodic character could not be physically maintained.

12. A simple pendulum, disturbed and then set free to oscillate under the constant action of gravity, soon comes to rest if allowed to communicate its motion to a surrounding medium: to maintain the oscillations, the pendulum requires to be repeatedly disturbed, and each impulse may change the phase and amplitude, and possibly also the direction of the vibration. In the same way, the vibrations of character at the points of a luminous body must be maintained by the repeated action of something analogous to an impulsive force. It is impossible to imagine that the representative impulse can always have the same magnitude and direction, and occur at the particular instant when the vibration is in a particular phase. Hence the vibration must, of almost absolute necessity, be different in its amplitude, phase, or direction, after every impulse.

Further, as already remarked in Art. 6, any luminous source available for experiment is not a geometrical point, but an area of appreciable magnitude, and the resultant effect at any point is due to the superposition of the effects of rays transmitted from every point of the luminous area: even if it were possible that the vibrations at a single point could be maintained constant in periodic character, it is inconceivable that the constancy of periodic character could be maintained at points belonging to an appreciable area.

A representative force.

13. In the case of a plane-polarised ray of constant character throughout the part considered, the vibratory motion of the representative point p is thus the same for all points P in the line of transmission, and only the phase of the vibration differs at different points at a given instant: hence the expression $y = a \sin \frac{2\pi}{\lambda} vt$, which represents the change of state at the time t at the point for which $\frac{2\pi x}{\lambda} - a = 0$, also represents the vibration at any other point of the ray, if we have due regard in every case to the epoch from

which the time is measured. The general expression for the law of the change at any point of a plane-polarised ray has been deduced on the assumption that the variation of the state with the time is exactly the same as the variation of position of an isochronous pendulum; or, what is the same thing, of a particle of unit mass vibrating in a straight line and attracted towards an origin in the line by a force of which the magnitude is proportional to the distance therefrom. For the velocity u of the attracted particle at the time t being $\frac{dy}{dt}$, the accelerative force at the same instant is $\frac{du}{dt}$ or $\frac{d^2y}{dt^2}$: by hypothesis the force is attractive, and is measured by f^2 times the distance, or by $-f^2y$, where f is a constant quantity: hence $\frac{d^2y}{dt^2} = -f^2y$.

It is easily seen that $y = B\sin(ft + \beta)$, in which B and β are both independent of the time, is a solution of this differential equation: for differentiating once we have $\frac{dy}{dt} = fB\cos(ft+\beta)$, and, differentiating a second time, $\frac{d^2y}{dt^2} = -f^2 B\sin(ft+\beta) = -f^2y$.

If the time be measured from an epoch of passage through the origin, the constant β is zero and the expression becomes $y = B\sin ft$.

Hence in the case of plane-polarised light, the vibratory motion of the representative point p, being expressed by the relation $y = a\sin\frac{2\pi}{\lambda}vt$, is identical with that of a particle of unit mass attracted towards the origin by a force which is measured by $\frac{4\pi^2 v^2}{\lambda^2}$ times the distance.

Even if the actual change of state at the point P were an oscillatory rotation of an ethereal particle about a diameter, as suggested by Rankine,[1] the above kinematical representation would still hold: in that case, the direction of the line Pp would represent that of the axis of rotation of the ethereal particle at P, and the distance Pp would represent the angular disturbance at the given instant.

Or again, the real change may be an electro-magnetic disturbance, whatever that may be.

The representative force is dependent on the luminous source.

14. But it will be obvious on reflection that the relation between the

---

[1] *Philos. Magazine*; 1853, ser. 4, vol. 6, p. 403.

distance of the ideal particle, and the ideal force which acting upon the ideal particle would cause a vibration isochronous with that of the periodic change involved in the transmission of the given ray of light, is generally independent of the specific properties of the transmitting medium. The ratio being $4\pi^2 r^2 : \lambda^2$ depends only on the ratio $\lambda : r$, that is to say, on the period of the vibration or the colour of the light. Now *simple* light generally retains its colour after transmission through any number of different media; it is only in fluorescent bodies that the colour of the light or the period of the change suffers alteration: whence we must infer that the period of vibration at any point of a ray, and thus the ratio of the ideal force to the distance, depends in general, not on the specific properties of the medium, but on the period of vibration of the change at the luminous source. The change of colour frequently observed after the passage of light through a medium is really due to the heterogeneity of the colour of the original light, and to the change of relative *intensity* (not period of vibration, or colour) of the component simple rays.

Further analogy with sound.

15. The same is true in the case of sound. Here, again, the transmission of a simple note causes a periodic change which may be represented algebraically by the same expression $y = a \sin \left\{ \dfrac{2\pi}{\lambda}(vt - r) + a \right\}$, and kinematically by the same to-and-fro motion of a particle attracted to an origin with a force measured by $\dfrac{4\pi^2 v^2}{\lambda^2}$ times the distance: and the constant ratio $4\pi^2 r^2 : \lambda^2$ depends only on the period of the vibration or the note of the sound, and thus on the source of the sound, not on the properties of the transmitting medium. Now the actual change of state at any point of a line of transmission of sound is known to be generally a to-and-fro motion of a particle of the medium, and the ideal particle and its motion may generally be taken to coincide with the real particle and its motion. But the magnitude of the representative force which acts on the ideal particle must not be confused with that of the elastic force which is evoked at the same point by the disturbance of the sound-transmitting medium: the representative force depends on the period of vibration at the source; the elastic force evoked by a given displacement depends on the specific properties of the medium: the resultant force acting on the real particle depends, not only on the specific properties of the medium, but on the continued action of the vibrating source.

The representative force in the case of the vibration of an elastic ether of which the effective density depends on the direction of the vibration.

16. To take another example: in the latest hypothesis as to the properties of an elastic luminiferous ether, it is assumed that the actual and effective elasticity of both volume and figure and the actual density of the ether are the same for all directions in a biaxal crystal, but that the effective density varies with the direction of vibration and is related to three mutually perpendicular lines. Hence, if the ether vibrates freely after disturbance parallel to one or other of these lines, the period of vibration will depend on the direction of the disturbance; for, though the effective elasticity is the same for each direction, the effective density, or effective mass to be put in motion, is different: and the ideal force, which acting on an ideal particle of unit mass gives a synchronous representative vibration, will have a different relation to the distance for the three directions of disturbance, although the ethereal elasticity, both actual and effective, is assumed to be really identical for all directions.

A fallacy.

17. We are now in a position to recognise the fallacy of a method which has been used for the derivation of Fresnel's wave-surface from the properties of an incompressible elastic ether. It is first proved that the elastic force evoked by a unit displacement along a line OP, which is the radius vector of a certain ellipsoid, if resolved along the direction of displacement OP, is $\frac{1}{OP^2}$: that if OP is an axis of a section of the ellipsoid, the other component is perpendicular not merely to OP but also to the plane of the section: that if the section has the direction of the wave-front, the second component is without effect owing to the incompressibility of the ether: that the effective elastic force for unit displacement is thus $\frac{1}{OP^2}$. It is then tacitly assumed that the effective elastic force is identical with the above representative force (which is measured by $\frac{4\pi^2 r^2}{\lambda^2}$ times the distance): hence it is inferred that $\frac{4\pi^2 c^2}{\lambda^2} = \frac{1}{OP^2}$. It is next wrongly assumed that *the wave-length λ is always the same for rays of the same colour transmitted in the same medium*, and that λ in the above relation is thus a constant: whence it is concluded that the velocity varies inversely as OP. That the proof is fallacious is clear from the last Article, in which it has been shown that, in the representative vibration, the relation of the ideal force to the ideal distance is independent of the specific properties of the medium and depends on the luminous source. Indeed the assumption of the constancy of λ is inconsistent with the conclusion, namely that v differs with the direction of vibration; it is obvious that the period $\frac{\lambda}{v}$ and the wave-length λ cannot be both constant if v be variable: the colour really depends on the *period of vibration*, not solely on the wave-length.

Fresnel himself proceeded in a different way, and assumed a relation founded on the analogy of a line of vibrating ethereal particles to a vibrating string.

In general, if a plane-polarised ray is transmissible in a given direction, the plane of polarisation can have at most two different positions.

18. We have seen (Art. 8) that if two plane-polarised rays of the same wave-length can be transmitted along the same line with the same velocity but with different positions of the plane of polarisation, they may be identical in effect with a single plane-polarised ray transmitted along the line with the same velocity but with an intermediate position of the plane of polarisation; the direction of the latter being determined by the ratio of the amplitudes of vibration of the component rays: conversely, the effect of a single ray of given plane of polarisation and simple colour is identical with that of two rays of the same simple colour transmitted along the same line with the same velocity, and with their planes of polarisation in any assigned positions.

Now a plane-polarised ray *can* be transmitted along the line of intersection of two planes of physical symmetry of the medium, for the planes of symmetry of the plane-polarised ray and the planes of symmetry of the medium may be taken to coincide: but the velocity of the ray will depend upon the position of the plane of polarisation, if the physical relations of the medium relative to the two planes of symmetry are different. If the latter be the case, as for instance when the line is an axis of symmetry of an ortho-rhombic crystal, no ray having a plane of polarisation oblique to the symmetral planes of the crystal can be transmitted along it: for such a ray, if transmissible, would be kinematically equivalent to two rays transmitted along the line with the *same* velocity, each having its plane of polarisation coincident with a different plane of symmetry; two rays can be actually transmitted with these positions of the plane of polarisation, but that their velocity should be equal is in general physically impossible.

In exactly the same way it follows that if along any line, whether an axis of symmetry or not, a plane-polarised ray can be transmitted with its plane of polarisation in two different positions but with different velocities in the two cases, a third position of the plane of polarisation is physically impossible.

The refraction of the medium cannot be higher than double.

19. Hence, for a given direction of transmission in such a medium, a plane-polarised ray cannot have more than two different velocities: and the medium cannot present more than double refraction; for, according to the undulatory theory, whatever the nature of the physical change, the direction of the refracted ray is dependent upon its velocity.

Degree of the equation of the ray-surface.

20. A diameter of the ray-surface for such a medium will thus intersect the surface in at most four real points, two on each side of the origin; and the equation of the ray-surface cannot be of a degree higher than the fourth, if it be granted that the above method of proof excludes the existence of imaginary velocities and imaginary points of intersection of a real line with the surface.

In fact, even if there be two imaginary positions of the plane of polarisation for a given real direction of ray-transmission, the imaginary velocities must be in general unequal, since the two planes will be differently related to the crystal and will thus correspond to different crystalline properties, whether real or imaginary. But even if the planes of polarisation be imaginary, the *difference* of the imaginary velocities of the two plane-polarised rays prevents the resultant effect from being that of a single plane-polarised ray with an imaginary plane of polarisation (Art. 8*b*).

The transmissibility of even a single plane-polarised ray is not a physical necessity: but if one position of a plane of polarisation be possible, there is a second at right angles with the first.

21. We may remark that it is not a physical necessity that a plane-polarised ray should be transmissible at all: a plane-polarised ray cannot be transmitted, for instance, along the morphological axis of a crystal of quartz.

As a plane-polarised ray is symmetrical to two planes, the plane of polarisation and the transverse plane, it would seem that if the characters of a crystal admit of one symmetral plane of the ray having a given position, they must admit of the other symmetral plane having the same position: in other words, for a given direction of transmission, if there is one possible position of the plane of polarisation, there is a second at right angles to the first. The same result is later arrived at in another way and the positions of the perpendicular planes are determined (Art. 40c).

Transmission of a ray along an axis of tetragonal or hexagonal symmetry.

22. On the other hand, the morphological axis of a tetragonal or hexagonal crystal is the intersection of two or more symmetral planes for which the physical relations are identical: hence along such a line it *is* physically possible to transmit two rays having the same velocity and different planes of polarisation, and thus having a resultant effect identical with that of a single plane-polarised ray. The amplitudes of the component rays being arbitrary may be so adjusted that the equivalent single plane-polarised ray has any plane of polarisation whatever: and it follows that, along the morphological axis of a given tetragonal or hexagonal crystal, a ray may be transmitted with any direction of the plane of polarisation, but in each case with the same velocity.

The velocity-factor.

23. The velocity of transmission of a plane-polarised ray of given colour is found to depend on the properties of the medium: since the vibration is in only a single direction, we may assume that the velocity of transmission corresponding to a given direction of vibration depends *solely* on the properties of the medium relative to the direction of the vibration. To avoid confusion of ideas, let the action of the medium, in so far as it affects the velocity of a ray of given direction of vibration, be said to be due to a *velocity-factor*; the magnitude of the factor depending on the properties of the medium for the direction of the periodic change or vibration.

The velocity-factor is necessarily the same for all directions perpendicular to an axis of tetragonal or hexagonal symmetry.

24. We have shown, from principles of mere symmetry of the medium and superposition of changes, without regard to their physical character, that along the morphological axis of a tetragonal or hexagonal crystal a ray is transmissible with the plane of polarisation in any azimuth whatever, and that the velocity of transmission of the ray is always the same: hence, for all directions of vibration perpendicular to the morphological axis of a uniaxal crystal, the velocity-factor has the same magnitude. It follows that the symmetry of the velocity-factor, at any rate for directions of

rectilinear vibration lying in a plane perpendicular to the tetragonal or hexagonal axis of symmetry, is of a higher order than that of the morphological development.

The corresponding geometrical character is worthy of remark, namely, that in a parallelepipedal system of points every plane of the system passing through an axis of tetragonal or hexagonal symmetry is a plane of symmetry for the planes and lines, though not for the points, of the system: the symmetry of the system relative to such a plane being in general "symmetry of aspect," and not absolute.[1]

Transmission of a ray in a direction lying in a plane of general symmetry but oblique to an axis of tetragonal or hexagonal symmetry.

25. Consider the case of a ray transmitted in one of the planes of symmetry S of a tetragonal crystal, but in a direction oblique to the morphological axis. Either plane of symmetry of the plane-polarised ray may be taken to coincide with the plane of symmetry S of the crystal: this is confirmed by experiment, for these directions of the planes of polarisation of a ray are found to be physically possible. But if the plane of polarisation of the ray is coincident with the plane of symmetry S of the crystal, and the vibration is assumed to be perpendicular to the plane of polarisation, the vibration is perpendicular to the morphological axis, whatever the position of the ray in the plane: hence, according to the preceding Article, the velocity-factor, and therefore the velocity of the ray, will be the same for all ray-directions in this plane, and one curve of intersection of the ray-surface with the plane of symmetry S of the crystal will be a circle. On the other hand, if the plane of polarisation of the ray is normal to the plane of symmetry S of the crystal, the vibration will be in the same plane of symmetry S and in a direction oblique to the morphological axis: the physical characters belonging to the direction of the vibration, including the velocity-factor, will thus vary with the direction of the ray, and the velocity-curve corresponding to those rays of which the plane of polarisation is normal to the symmetral plane S of the crystal will not be circular: the curve will be symmetrical, however, both to the morphological axis and a line perpendicular to it, for they are directions with respect to which all the characters of the crystal are symmetrical. Further, the second curve will touch the first at its points of intersection with the morphological axis: for the two directions perpendicular to that line, and lying respectively in and perpendicular to the plane of symmetry S, are by

---

[1] H. J. S. Smith; *Philosophical Magazine*, 1877, ser. 5, vol. 4, p. 18.

hypothesis similar in all their relations, and correspond therefore to the same velocity-factor; hence both curves meet on the morphological axis, and therefore touch each other, for the morphological axis divides each curve symmetrically.

But since the equation of the ray-surface has been shown to be of a degree not higher than the fourth, and the equation of one curve of intersection, a circle, is of the second degree, that of the other curve will likewise be of the second degree, and therefore represent an ellipse — for the curve is closed and has unequal diameters. This result agrees with the experimental discovery made by Huygens.

Transmission of rays along the axes of symmetry of an ortho-rhombic crystal.

26. Take next the case of an ortho-rhombic crystal. In the first place, as shown in Art. 18, a ray can be transmitted along any of the axes of symmetry, and have its plane of polarisation coincident with either of the symmetral planes of the crystal which intersect therein. The three axes of symmetry being independent of each other in all their physical relations, the velocity-factors will be independent; and vibrations parallel to the several axes will thus in general correspond to different velocities of transmission. Let the velocity corresponding to an axis OX, OY, or OZ, *considered as a direction of vibration*, be denoted by a, b, or c respectively: then two rays are transmissible along OX with velocities b and c, and planes of polarisation normal to OY and OZ respectively: two rays are transmissible along OY with velocities c and a, and planes of polarisation normal to OZ and OX respectively: two rays are transmissible along OZ with velocities a and b, and planes of polarisation normal to OX and OY respectively.

Transmission of rays in a symmetral plane of an ortho-rhombic crystal.

27. Again, as far as directions lying in the plane of symmetry OXZ are concerned, there is no essential difference between an ortho-rhombic and a tetragonal crystal, if OZ is the morphological axis of the latter. The essential difference between two such crystals is that in one of them (the ortho-rhombic) the third axis of symmetry OY is independent of OX in its physical relations, and in the other (the tetragonal) is identical therewith. Hence we may infer that in the symmetral plane OXZ of an ortho-rhombic crystal, a ray is transmissible in any direction with its plane of polarisation either coincident with or perpendicular to that plane. Also, as in the case of a tetragonal crystal, the intersection of the ray-

surface with the symmetral plane will be a circle and a concentric ellipse: but, in the ortho-rhombic crystal, the circle and ellipse will be independent of each other in magnitude, since the velocity-factor for the direction of vibration OY is independent of those for the directions of vibration OX, OZ.

Intersections of the ray-surface with the symmetral planes of an ortho-rhombic crystal.

28. The intersections of the ray-surface with the axial planes OYZ, OZX, OXY of an ortho-rhombic crystal will thus be given by the following equations:—

$$(y^2+z^2-a^2)(b^2y^2+c^2z^2-b^2c^2)=0,$$
$$(z^2+x^2-b^2)(c^2z^2+a^2x^2-c^2a^2)=0,$$
$$(x^2+y^2-c^2)(a^2x^2+b^2y^2-a^2b^2)=0.$$

General equation of the ray-surface for an ortho-rhombic crystal.

29. The equation of the ray-surface itself must be of the form
$$(y^2+z^2-a^2)(b^2y^2+c^2z^2-b^2c^2)+x\,\phi(xyz)=0,$$
since it reduces to the first expression when x is made zero. But according to Art. **20** the quantity $x\,\phi(xyz)$ cannot consist of terms of degrees higher than the fourth: further, the surface being symmetrical to the axial planes, its equation can only involve even powers of x, y, z: hence the only terms which can enter the expression $x\,\phi(xyz)$ are x^4, z^2x^2, x^2y^2 and x^2.

The general equation is thus of the form
$$(y^2+z^2-a^2)(b^2y^2+c^2z^2-b^2c^2)+A x^4+B z^2 x^2+C x^2 y^2+D x^2=0;$$
or, multiplying out,
$$A x^4+b^2y^4+c^2z^4+(b^2+c^2)y^2z^2+B z^2 x^2+C x^2 y^2+D x^2-b^2(c^2+a^2)y^2$$
$$-c^2(a^2+b^2)z^2+a^2b^2c^2=0.$$

Also, it is evident from the equations of the curves of intersection with the three axial planes that x, y, z, and a, b, c, are simultaneously cyclically interchangeable (Art. **28**); hence
$$A=a^2;\quad B=c^2+a^2;\quad C=a^2+b^2;\quad D=-a^2(b^2+c^2).$$
Substituting these values, the equation becomes
$$(x^2+y^2+z^2)(a^2x^2+b^2y^2+c^2z^2)-a^2(b^2+c^2)x^2-b^2(c^2+a^2)y^2-c^2(a^2+b^2)z^2+a^2b^2c^2=0$$
or, multiplying by r^2,
$$r^2(a^2x^2+b^2y^2+c^2z^2)-r^2\{a^2(b^2+c^2)x^2+b^2(c^2+a^2)y^2+c^2(a^2+b^2)z^2\}+a^2b^2c^2(x^2+y^2+z^2)=0$$
or $\quad a^2x^2(r^2-b^2)(r^2-c^2)+b^2y^2(r^2-c^2)(r^2-a^2)+c^2z^2(r^2-a^2)(r^2-b^2)=0$
$$\text{or}\quad \frac{a^2x^2}{r^2-a^2}+\frac{b^2y^2}{r^2-b^2}+\frac{c^2z^2}{r^2-c^2}=0:$$
which is Fresnel's equation of the ray-surface.

The ray-surface for a mono-symmetric or anorthic crystal.

30. (*a*) Next consider the case of a crystal which admits of the transmission of a ray of plane-polarised light in any direction, but presents only a single plane of geometrical and physical symmetry, and thus belongs to the mono-symmetric system: let the normal of the plane of symmetry be OY.

Since, from a purely geometrical point of view, a mono-symmetric crystal may be regarded as a homographic transformation of an ortho-rhombic crystal, it first suggests itself that the ray-surface for a mono-symmetric crystal may be such as would result from a corresponding transformation of the ray-surface for an ortho-rhombic crystal. That the analogy is imperfect, however, is evident from the fact that there is no corresponding distortion of the planes of polarisation; whatever the direction of ray-transmission within the mono-symmetric crystal, the planes of polarisation of the two transmissible rays are perpendicular to each other (Art. 21).

1. As in Art. 27, any ray whatever lying in the plane of symmetry can have that plane for either its plane of polarisation or its transverse plane: hence, exactly in the same way as before, it follows that the plane of symmetry intersects the ray-surface in two curves, the one a circle, the other a concentric ellipse: the former corresponding to the rays which have the symmetral plane for the plane of polarisation, the latter to the rays for which the plane of symmetry is the transverse plane. If OX, OZ, be the axes of the ellipse, a ray transmitted along OX will thus have its representative vibrations parallel to either OY or OZ; and a ray transmitted along OZ will have its vibrations parallel to either OY or OX.

2. A ray transmissible along the line OY can have its plane of polarisation in only one or other of two directions of which the normals are perpendicular both to each other and to the line OY (Art. 21).

3. Since the elliptic and circular sections of the ray-surface made by the plane XOZ are both of them symmetrical to the lines OX, OZ, while the plane XOZ is a plane of general physical symmetry of the crystal, and its normal OY is an axis of general symmetry of diagonal type, we may reasonably assume that *for this particular property* (so long as there is no variation of colour or temperature) the planes YOX, YOZ, are themselves planes of symmetry of the crystal; in which case, the lines OX, OZ, will be the directions of vibration of the two rays transmissible along the axis OY.

4. For the given colour and temperature, the circumstances are identical, for this particular property, with those of an ortho-rhombic crystal having OX, OY, OZ, for axes of symmetry: and the ray-surface

will thus for a mono-symmetric crystal have the same general form as for an ortho-rhombic one.

(*b*) The general form of the ray-surface, being quite unaffected by the degradation of the symmetry from the ortho-rhombic to the mono-symmetric type, is clearly independent of the type of symmetry altogether: the general form will therefore be the same even for an anorthic crystal.

The difference in the type of symmetry thus affects, not the general form of the ray-surface, but only the constancy of the directions and relative lengths of the axes of the surface for different colours and temperatures. An axis of general symmetry of the crystal is necessarily an axis of symmetry of the ray-surface whatever the colour of the light or the temperature of the crystal (p. 21).

The form of the ray-surface is independent of the physical character of the periodic change.

31. The rigorous accuracy of the form assigned to the ray-surface by Fresnel is thus a necessary consequence of the general features of perpendicularly transverse vibrations, independently of the physical character of the change.

And although, as in the case of an incompressible elastic ether with effective rigidity dependent upon the direction of vibration, the same form of ray-surface may result notwithstanding the obliquity of the transverse vibration, this is not generally true. The form of the ray-surface which follows, for example, from a version of the elastic theory of double refraction suggested by Rankine and further developed by Lord Rayleigh is different from that of Fresnel, and only gives the latter as a first approximation. That version, according to which the ether is incompressible and has an effective density dependent on the direction of vibration, involves the general obliquity of the latter to the direction of transmission.[1]

In fact, whatever the degree of symmetry of the characters of a plane-polarised ray as transmitted within a medium, the above form of ray-surface will result from any hypothesis which has for necessary consequence that if one plane-polarised ray is transmissible in a given direction, a second plane-polarised ray is transmissible in the same direction with a different velocity and has its plane of polarisation perpendicular to that of the first.

It may be remarked that, in the above reasoning, no assumption as to the molecular constitution of the ether has been necessary.

---

[1] *Philosophical Magazine;* 1851, ser. 4, vol. 1, p. 441: 1888, ser. 5, vol. 26, pp. 525, 527.

Resilience.

32. It was explained in Art. 14 that the period of the change at any point of a ray of light depends in general on the period of the change at the luminous source, and not on the specific properties of the medium; but the latter may conceivably affect some or all of the remaining characters of the ray, namely, amplitude and direction of vibration, velocity and direction of transmission through the medium. The property by virtue of which a periodic change of any kind is transmissible through a medium may be denoted by the general term *resilience*: we may imagine that a disturbance at any point of the medium evokes an opposing resilience of which the magnitude increases with the amount of the disturbance. Optical resilience, in so far as it affects only the velocity of transmission of a periodic change having a given direction, is identical with the velocity-factor for that direction, mentioned in Art. 23. When the periodic change is a vibratory motion such as follows the removal of a compressing or distorting force, resilience is identical with elasticity of volume or figure.

From this point of view, the periodic change at any point of a ray of plane-polarised light may be treated as a resultant effect of two forces; the one an *initiatory* linear force periodic in its variations, and having a period identical with that of the luminous source; the other a *secondary* force or a resilience, evoked by the disturbance produced by the initiatory force.

Free and forced vibrations.

33. The periodic change at a point of a ray of light is a *forced* vibration, resulting from the continued action of the luminous source: it differs from a *free* vibration, such as would be produced by resilience alone if the luminous source were removed while the medium is in a state of disturbance.

A simple case of free vibration.

34. In the simplest possible case of free vibration of a character of a medium, we may imagine that the disturbance at the point is of such a kind that at any instant it can be represented by the length and direction of a straight line y drawn from an ideal particle of unit mass to the point, and that the resilience of the medium can be represented by an ideal force acting in the line of disturbance, tending to diminish the disturbance, and proportional in magnitude to the disturbance itself, the proportion being independent of direction: such a medium may be said to be *isotropically* resilient for the given character.

As in Art. 13 we may write
$$\frac{d^2y}{dt^2} = -f^2 y,$$
where f is independent of the time and depends on the properties of the medium.

A solution of this equation is $y = B\sin(ft+\beta)$, where B and β are constants: the expression represents a vibration of which the period is $\frac{2\pi}{f}$, since any value of y recurs when t is increased by an integral multiple of that quantity.

As already pointed out, such a mode of representation is still possible, even when the actual change is an oscillatory rotation of an ethereal particle (Art. 13).

A simple case of forced vibration.

35. But suppose that in the above medium the vibration at the point is not free but forced, and that the initiatory force is a periodic one related to the time in the same way as the disturbance at a point of a plane-polarised ray of simple colour; the initiatory force can in such case be represented by an expression of the form $S\sin st$, where S and s are constants, and the latter depends only on the period of vibration of the luminous source. As before, the ideal resultant force acting on the ideal particle of unit mass is $\frac{d^2y}{dt^2}$, and is due to the superposition of the initiatory force $S\sin st$ and the resilience $-f^2y$: hence $\frac{d^2y}{dt^2} = S\sin st - f^2 y$.

It is easily seen that $y = B\sin st$ is a solution of this differential equation: for differentiating, we get first $\frac{dy}{dt} = Bs\cos st$, and next $\frac{d^2y}{dt^2} = -Bs^2 \sin st$: substituting in the above equation, and dividing by $\sin st$, we get $B = \frac{S}{f^2 - s^2}$: and thus $y = \frac{S}{f^2 - s^2} \sin st$. Hence the resilience affects merely the amplitude, not the period or general character of the vibration at the point.

The resilience, being $-f^2 y$, has likewise the same period as the initiatory force.

Transmission of a simple forced vibration in an isotropically resilient medium.

36. If a luminous source is in a state of periodic vibration represented

kinematically by the linear motion of a particle attracted to an origin by a force proportional to the distance, and is surrounded by a medium such that the resilience is represented by a force acting in the line of disturbance and proportional to it in magnitude, the changes transmitted through the medium along a given direction perpendicular to that of the vibration may thus be expected to be always in the same plane and have the same period; no resilient force oblique to the plane containing the direction of ray-transmission and the direction of vibration of the luminous source is evoked by the disturbance: in any direction in an isotropically resilient medium, a plane-polarised ray, if transmissible at all, may thus be transmitted with any azimuth of plane of polarisation whatever.

A more general case of free vibration of an æolotropically resilient medium.

37. As a more general case, we may imagine that in a crystalline medium there are three directions, not co-planar, inclined obliquely or perpendicularly to each other, for each of which a disturbance evokes a resilience which in its effects is represented by an ideal force, contrary and proportional to the disturbance, acting on an ideal particle of unit mass; the relation of the ideal representative force to the distance of an ideal attracted particle of unit mass being, however, like most other physical characters, different for the three directions: the latter may be termed *axes of optical resilience*.

That such a representation is possible, even in an elastic ether of which the elasticity is the same in all directions, has already been pointed out in Art. 16: for if the effective density depends on the direction of vibration, the period of a free vibration will also vary with the direction, since although the real accelerative force has the same constant relation to the distance it will have a different effective mass to keep in motion.

When it is desirable to emphasise the fact that the resilient force under consideration is the ideal force which would produce an analogous to-and-fro motion of a particle of unit mass and not the statical force necessary to the maintenance of a given state of disturbance, we may conveniently distinguish it as *vibrational resilience*. If the three constants of vibrational resilience be respectively e^2, f^2, g^2, and x, y, z, be the distances which represent the disturbances parallel to the respective axes at any time t, we have for a free vibration due to a disturbance along each of the axes

$$\frac{d^2x}{dt^2} = -e^2 x\ ;\ \frac{d^2y}{dt^2} = -f^2 y\ ;\ \frac{d^2z}{dt^2} = -g^2 z:$$

whence, in the same way as before,

$$x = A\sin(et + \alpha)\ ;\ y = B\sin(ft + \beta)\ ;\ z = C\sin(gt + \gamma).$$

According to the principle of superposition of changes, if the direction of the initial disturbance at the point is inclined to the three axes of resilience, the initial disturbance may be resolved along those directions, and the resultant free vibration is such as would result from the composition of the free vibrations corresponding to the several axial directions. Hence, if the vibration is free, the disturbance at a given instant is determined by the above triad of equations.

Since the ratios $x : y : z$ depend on the time, the motion of the representative particle is not in a straight line passing through the origin. The particle, in fact, describes a curve in three dimensions, and never passes twice through the same position unless the ratios $e : f : g$ are commensurable.

The quantities e^2, f^2, g^2, may be conveniently termed *coefficients of optical vibrational resilience*: and the medium may be said to be *æolotropically* resilient. The coefficients of vibrational resilience are independent of x and t for the same ray, but even with the same medium may conceivably be different for different rays, and thus vary with the period of the change, or in other words, with the colour of the light.

In Art. 42 it is pointed out that obliquity of mutual inclination of the axes of optical resilience is not met with even in mono-symmetric or anorthic crystals.

A more general case of forced vibration of an æolotropically resilient medium.

38. Consider next a forced vibration of a crystalline medium having three dissimilar oblique or rectangular axes of vibrational resilience as before: assume that the initiatory force at any point of a ray may again be represented by an expression of the form $S\sin st$, where s is a constant depending on the period of the change at the luminous source.

If OP be any line passing through an origin O, and OL, OM, ON, lengths measured along the axes of resilience, be edges of a parallelepiped of which OP is a diagonal, whatever the length OP we have

$$OL = \lambda \cdot OP \quad OM = \mu \cdot OP \quad ON = \nu \cdot OP,$$

where $\lambda\ \mu\ \nu$ are constants for a given direction of OP.

From the principle of superposition, it follows that the initiatory force $S\sin st$ acting in the line OP can be resolved into three initiatory forces $\lambda S\sin st,\ \mu S\sin st,\ \nu S\sin st$, acting along the axes of $x\ y\ z$ respectively.

In exactly the same way as before we have the following expressions for the several vibrations parallel to the respective axes:—

$$x = \frac{\lambda S}{e^2 - s^2}\sin st; \quad y = \frac{\mu S}{f^2 - s^2}\sin st; \quad z = \frac{\nu S}{g^2 - s^2}\sin st;$$

where the quantities $\lambda\ \mu\ \nu\ e^2\ f^2\ g^2\ S$ and s are all independent of the time.

Hence the ratios $x : y : z$ are also independent of the time, and the representative particle vibrates in a straight line through the origin.

The period of the resultant vibration is identical with that of the initiatory force, but the direction of the vibration is different. If $\lambda'\ \mu'\ \nu'$ determine the direction of the resultant vibration,

$$\lambda' : \mu' : \nu' = x : y : z = \frac{\lambda}{e^2 - s^2} : \frac{\mu}{f^2 - s^2} : \frac{\nu}{g^2 - s^2}.$$

Further, the axial components of the representative resilient force being $-e^2 x,\ -f^2 y,\ -g^2 z$, or

$$-\frac{e^2 \lambda S}{e^2 - s^2} \sin st,\quad -\frac{f^2 \mu S}{f^2 - s^2} \sin st,\quad -\frac{g^2 \nu S}{g^2 - s^2} \sin st,$$

the resultant resilient force will have a direction determined by the ratios

$$\frac{e^2 \lambda}{e^2 - s^2} : \frac{f^2 \mu}{f^2 - s^2} : \frac{g^2 \nu}{g^2 - s^2}.$$

Hence the resultant resilient force always acts in the same direction throughout the vibration, but it is inclined to the direction $\lambda\ \mu\ \nu$ of the initiatory force and also to the line of vibration $\lambda'\mu'\nu'$, both of which pass through the origin: further, the resultant resilient force has the same period as the initiatory force.

Transmission of a simple forced vibration in an æolotropically resilient medium.

39. In such a medium, therefore, an initiatory linear periodic force having a direction inclined to an axis of resilience and acting at a given point gives rise at that point to a linear periodic vibration in a direction inclined to the initiatory force, and to a resilience of which the resultant effect is represented by a periodic force acting on the ideal particle in a third and constant direction not passing through the given point. Since the periodic change is transmitted through the medium by virtue of the resilience, and action is always equal and contrary to reaction, we should thus expect that, along any line of transmission, the direction of the periodic change will in general vary from point to point of the ray; and that the transmitted periodic change can only be in a direction lying always in the same plane, if the plane containing the initiatory force and the direction of transmission likewise contains the direction of the resilient force, and therefore also the direction of representative vibration.

Consider, for example, the case of a ray transmissible along an axis OX of an ortho-rhombic crystal : from the symmetry it follows that a plane-polarised ray transmitted along OX must have its vibrations parallel to one or other of the dissimilar axes OY, OZ, and that for a ray of given simple colour the velocity of transmission will depend on the direction of vibration. If, however, the initiatory force at the initial point of the ray, though perpendicular to OX, is oblique to the axes OY, OZ, it may be resolved into two forces, parallel to OY, OZ, respectively, and each may be regarded as originating a simple plane-polarised ray: the motion of the representative point will be the resultant of the motions belonging to each ray, and will thus be continually changing its direction as the disturbance is transmitted along OX.

Case of an ortho-rhombic crystal.

40. (*a.*) *Direction of the resultant vibrational resilience for a given disturbance.* For simplicity, let the crystalline medium present three mutually perpendicular but dissimilar symmetral planes, and thus belong to the ortho-rhombic system: the axes of resilience necessarily coincide with the crystallographic axes, the lines of intersection of the symmetral planes. Let X, Y, Z, be the components, parallel to the axes of co-ordinates, of the representative resilient force corresponding to a disturbance defined by the co-ordinates $x'\ y'\ z'$: then
$$X = -e^2 x'\ ;\ \ Y = -f^2 y'\ ;\ \ Z = -g^2 z'.$$
The direction-cosines of the resultant resilience F are in the ratios
$$X:\ Y:\ Z:\ \text{or } e^2 x':\ f^2 y':\ g^2 z'.$$
But if an ellipsoid $e^2 x^2 + f^2 y^2 + g^2 z^2 = 1$ be of such dimensions that it passes through the point $P\ (x' y' z')$, the direction-cosines of PG the normal of the ellipsoid at the point $x' y' z'$ are likewise in the ratios $e^2 x':\ f^2 y':\ g^2 z'$ (Fig. 19).

Hence the resultant resilient force due to a disturbance OP acts in the direction PG of the normal of the ellipsoid $e^2 x^2 + f^2 y^2 + g^2 z^2 = 1$ at the point $P\ (x' y' z')$ lying on its surface.

(*b.*) *Direction of transmission of a plane-polarised ray of which the direction of the plane of polarisation is given.*

If O be the centre of the ellipsoid and OP the representative direction of vibration, the initiatory force must also lie in the plane OPG which contains the direction of vibration and the secondary force: further, the direction of transmission must lie in the same plane (Art. **39**). The vibration being always perpendicularly transversal to the direction of transmission, the ray Or corresponding to the vibration OP is thus in the plane OPG and perpendicular to OP (Fig. 19).

(*c.*) *The planes of polarisation for a given direction of ray are perpen-*

dicular to each other, and have directions which can be defined by means of an ellipsoid.

It has already been proved (Chapter IV, Art. 24) that if OP be a central radius vector of an ellipsoid, and PG the normal of the ellipsoid at P, the line OP is an axis of the section of the ellipsoid by a plane through OP perpendicular to the plane OPG ; for a given direction of ray Or there are thus two possible directions of vibration OP_1, OP_2, which can be transmitted without change of plane, and they are the axes of the section of the ellipsoid by a plane to which the ray Or is normal. Hence the planes of polarisation corresponding to a given direction of ray are perpendicular to each other and are determined by the above geometrical construction.

(d.) *Magnitudes of the total and effective vibrational resilience for a given disturbance.*

If F be the resultant resilience, $F^2 = X^2 + Y^2 + Z^2 = e^4 x'^2 + f^4 y'^2 + g^4 z'^2$.

But if OM (Fig. 19) be the central normal to the tangent plane at P $(x' y' z')$,

$$\frac{1}{OM^2} = e^4 x'^2 + f^4 y'^2 + g^4 z'^2. \quad \text{(Chap. IV, Art. 4).}$$

Hence the resultant resilience F is measured by $\frac{1}{OM}$.

The resilience being in the direction PG, and the actual vibration in the direction PO, the effective resilience is $F \cos OPG$

$$= F \frac{OM}{OP} = \frac{1}{OP}.$$

This corresponds to a disturbance of magnitude OP: hence the effective resilience for a unit disturbance in the direction OP is $\frac{1}{OP^2}$.

(e.) *Relation between the effective vibrational resilience and the velocity of transmission.*

In the development of his theory of Double Refraction, Fresnel was compelled to make an assumption as to the relation between the effective elastic force and the velocity of normal-propagation of the corresponding wave, and supported his assumption by reference to the analogy of a vibrating string.

In the preceding Articles, all the forces are purely representative, and the assumptions and reasoning founded thereon are really independent of the physical character of the change. But it is clear that the velocity of transmission must depend on the physical character of the periodic change, and that it is impossible to proceed farther and deduce the absolute velocity of transmission without some assumption involving the nature of the change and the constitution of the ether. All that we have been able to suggest hitherto is that the velocity is in some way dependent on the characters of the medium relative to the direction of the vibration, and these characters have been collectively expressed by the term velocity-factor: in other words, it was suggested that the velocities of the two rays transmissible in the direction Or are determined by some function of the directions of vibration OP_1, OP_2, and thus by some function of the lengths OP_1, OP_2, for the length of a radius vector of an ellipsoid is determined by the direction. But we have also shown (Art. 29) that, without any assumption as to the real nature of the change, it is possible to determine the velocities r_1, r_2, of the two rays transmissible in a given direction, in terms of a, b, c, the velocities corresponding to vibrations in the directions of the principal axes: hence the velocity r of transmission along Or is necessarily so related to OP, an axis of the section of the ellipsoid $e^2 x^2 + f^2 y^2 + g^2 z^2 = 1$ by the plane perpendicular to Or, that the equation

$$\frac{a^2 x^2}{r^2 - a^2} + \frac{b^2 y^2}{r^2 - b^2} + \frac{c^2 z^2}{r^2 - c^2} = 0$$

represents the ray-surface. There is only one relation between r and OP which leads to this form of ray-surface, namely $r = OP$: we are thus compelled to infer that the velocity of transmission of a ray is directly proportional to that radius vector OP of the above ellipsoid $e^2 x^2 + f^2 y^2 + g^2 z^2 = 1$, which has the same direction as the vibrations of the ray: from Art. 40d it follows that the same relation is expressed by the statement that the ray-velocity is *inversely* proportional to the square root of the effective resilience for unit disturbance in the direction of vibration.

It follows from the above relation that:—

If a line Or is perpendicular to a central section of the ellipsoid $e^2 x^2 + f^2 y^2 + g^2 z^2 = 1$, and OP_1, OP_2, are the axes of the section, a plane-polarised ray can be transmitted along Or, having OP_1 or OP_2 for the normal of its plane of polarisation and a velocity of transmission measured by OP_1 or OP_2, respectively.

That this relation is consistent with the form of the ray-surface arrived at in Art. 29 may be proved as follows:—

(*f.*) *Transformation of the above construction.*

From O draw OM perpendicular to the plane which touches the ellipsoid $e^2 x^2 + f^2 y^2 + g^2 z^2 = 1$ at the point P, and let M be the foot of the perpendicular: produce OM to R, making $OR \cdot OM = 1$.

(1.) First—find the locus of the points R when P takes all positions on the ellipsoid $e^2 x^2 + f^2 y^2 + g^2 z^2 = 1$.

AN ORTHO-RHOMBIC CRYSTAL.

Let $x'y'z'$ be the co-ordinates of P (Fig. 19); the tangent plane at P is $e^2x'x+f^2y'y+g^2z'z = 1$: hence, as in Art. 40d,

$$\frac{1}{OM^2} = e^4x'^2+f^4y'^2+g^4z'^2.$$

If $\xi\,\eta\,\zeta$ be the co-ordinates of the point R, we have $\dfrac{\xi}{e^2x'} = \dfrac{\eta}{f^2y'} = \dfrac{\zeta}{g^2z'}$; for, by construction, the line OMR is perpendicular to the tangent plane $e^2x'x+f^2y'y+g^2z'z = 1$.

Each of these fractions is equal to $\dfrac{\sqrt{(\xi^2+\eta^2+\zeta^2)}}{\sqrt{(e^4x'^2+f^4y'^2+g^4z'^2)}}$; that is $OR \cdot OM$, or unity.

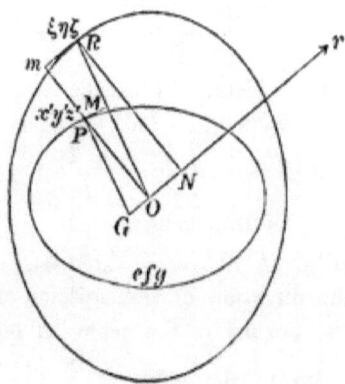

Fig. 19.

We thus have $\dfrac{\xi}{e} = ex'$; $\dfrac{\eta}{f} = fy'$; $\dfrac{\zeta}{g} = gz'$:

whence, since $e^2x'^2+f^2y'^2+g^2z'^2 = 1$,

it follows that $\dfrac{\xi^2}{e^2}+\dfrac{\eta^2}{f^2}+\dfrac{\zeta^2}{g^2} = 1.$

This is the equation of the locus of the points R, and represents an ellipsoid with the same symmetral planes as $e^2x^2+f^2y^2+g^2z^2 = 1$, but with semi-axes $e\,f\,g$ instead of $\dfrac{1}{e}\,\dfrac{1}{f}\,\dfrac{1}{g}$.

(2.) The equation of the plane which touches the ellipsoid $\dfrac{\xi^2}{e^2}+\dfrac{\eta^2}{f^2}+\dfrac{\zeta^2}{g^2} = 1$ at the point $R\,(\xi\,\eta\,\zeta)$ is

$$\frac{x\xi}{e^2}+\frac{y\eta}{f^2}+\frac{z\zeta}{g^2} = 1.$$

If Om be the perpendicular on this plane from the origin, the direction-cosines of Om have the ratios $\dfrac{\xi}{e^2} : \dfrac{\eta}{f^2} : \dfrac{\zeta}{g^2}$, or $x' : y' : z'$; hence the line Om passes through the point P.

RN, the normal of the ellipsoid $\dfrac{\xi^2}{e^2}+\dfrac{\eta^2}{f^2}+\dfrac{\zeta^2}{g^2}=1$ at R, is parallel to Om, and will thus intersect the line Or which lies in the plane ORP: if N be the point of intersection, $RN = Om$.

(3.) In the same way as before, since Om is the perpendicular from the origin to the plane $\dfrac{x\xi}{e^2}+\dfrac{y\eta}{f^2}+\dfrac{z\zeta}{g^2}=1$,

$$\frac{1}{RN^2} = \frac{1}{Om^2} = \frac{\xi^2}{e^4}+\frac{\eta^2}{f^4}+\frac{\zeta^2}{g^4} = x'^2 + y'^2 + z'^2 = OP^2.$$

The plane ORN is thus identical with the plane OPG; the normal RN to the ellipsoid at the point R has the same direction as OP and represents the direction of the vibration: $\dfrac{1}{RN}$ is equal to OP and therefore measures the velocity of transmission.

Hence the relation given above in Art. 40e is equivalent to the following:—If Or is the direction of transmission of a ray, the direction of the vibration, or the normal of the plane of polarisation, is normal to the line Or and also to the ellipsoid $\dfrac{x^2}{e^2}+\dfrac{y^2}{f^2}+\dfrac{z^2}{g^2}=1$; its velocity is measured by the *inverse* of the length of this normal intercepted between the ray and the ellipsoid: further the normal of the ellipsoid is perpendicular to the plane of polarisation of the corresponding ray.

It has been proved in Chapter IV that the ray-surface which follows from this relation is

$$\frac{\frac{1}{e^2}x^2}{r^2-\frac{1}{e^2}} + \frac{\frac{1}{f^2}y^2}{r^2-\frac{1}{f^2}} + \frac{\frac{1}{g^2}z^2}{r^2-\frac{1}{g^2}} = 0;$$

it has also been proved (Art. 29), without any assumption as to the real nature of the periodic change, that the equation of the ray-surface is

$$\frac{a^2 x^2}{r^2-a^2} + \frac{b^2 y^2}{r^2-b^2} + \frac{c^2 z^2}{r^2-c^2} = 0:$$

the results are consistent with each other if $e = \dfrac{1}{a}$, $f = \dfrac{1}{b}$, $g = \dfrac{1}{c}$; the

ellipsoid $\frac{x^2}{e^2}+\frac{y^2}{f^2}+\frac{z^2}{g^2}=1$ is identical with the optical indicatrix
$$a^2x^2+b^2y^2+c^2z^2=1.$$

Comparison with Fresnel's elastic forces.

41. (α.) If a, b, c, are the velocities of transmission of those rays of which the vibrations are parallel to the axes of x, y, z, respectively, Fresnel's method of derivation requires the elastic forces evoked by unit displacements along the axes to be taken as a^2, b^2, c^2, respectively: according to the above method, the ideal forces, which by their action on an ideal particle of unit mass would produce vibrations synchronous with those of the medium, will be measured by e^2, f^2, g^2, or $\frac{1}{a^2}, \frac{1}{b^2}, \frac{1}{c^2}$, for unit displacements of the ideal particle along those directions.

(β.) In Fresnel's method, the evoked elastic force normal to the direction of vibration of a real particle of ether is regarded as of no effect owing to the incompressibility of the medium: in the above method, no assumption is made as to the compressibility or incompressibility of the medium, but that component of the representative resilient force which is normal to the direction of vibration of an ideal particle is regarded as balanced by an equal component of the representative initiatory force at the same point of the ray.

Case of a mono-symmetric or anorthic crystal.

42. If the medium could present three dissimilar axes of optical resilience obliquely inclined to each other, it would follow as before that the axial components of the resilient force, corresponding to a disturbance defined by the co-ordinates $x'y'z'$, would be
$$X=-e^2x'\ ;\ Y=-f^2y'\ ;\ Z=-g^2z':$$
but the resultant resilient force F would no longer act along the normal to the ellipsoid $e^2x^2+f^2y^2+g^2z^2=1$, and the planes of polarisation of the rays transmissible along a given direction would no longer be at right angles to each other.

As such a character is not presented by any crystal which has been examined, we may infer that in all crystalline media the axes of optical resilience for a given colour and temperature are always mutually perpendicular, and that the symmetry of the crystal merely affects the directions of the triads of perpendicular lines and the ratios of the corresponding coefficients of optical resilience.

An unsatisfactory variation of Fresnel's method.

43. At first sight it would seem that the following would be a satisfactory and more simple mode of altering Fresnel's assumptions and reasoning so as to accord with the recent conclusion that the vibration is parallel, not to the radius vector RO, but to the normal RN of the ellipsoid $a^2x^2 + b^2y^2 + c^2z^2 = 1$ (Fig. 19).

Let $-a^2\xi, -b^2\eta, -c^2\zeta$, be the resolved axial components of the elastic force on a particle, due to a displacement from the centre O to the point R (ξ η ζ) lying on the surface of the ellipsoid $a^2x^2+b^2y^2+c^2z^2=1$. It may be shown as above that the elastic force acting on the particle when at R is directed along the normal RN and measured by $\frac{1}{RN}$, N being as usual the point of intersection of the normal with the ray: hence, the elastic force *for unit displacement measured perpendicularly to the ray*, and thus parallel to the direction of vibration, is $\frac{1}{RN^2}$.

Hence, if the particle were set free after having been displaced to the point R and no other force than the evoked elasticity were acting upon it, the *initial* motion would be along the normal RN under the action of a force which is measured by $\frac{1}{RN^2}$ times the distance from the ray. Assuming that the vibration is actually and permanently perpendicular to the ray, a periodic constraining force is requisite to maintain the isochronous character of the motion: if it were possible that the constraining force and the evoked elasticity could together be always measured by $\frac{1}{RN^2}$ times the distance of the particle from the direction of the ray, we might infer by analogy with the case of sound that the velocity of transmission would be measured by $\frac{1}{RN}$, for $\frac{1}{RN}$ is the square root of the effective elastic force due to a unit displacement in the direction of vibration.

It will be found, however, that such a motion cannot actually take place: the particle will only vibrate rectilinearly if the path passes through the origin. If the particle is not moving in a line through the origin, the evoked elastic force will be constantly changing direction; for at any instant it acts parallel to the normals of the ellipsoid $a^2x^2+b^2y^2+c^2z^2=1$ at the points where a line joining the particle to the centre O meets the surface. Further, the initiatory periodic constraining force is zero, not when the particle is at its position of maximum displacement, but when it is in its position of no disturbance.

The transmission of elliptically or circularly polarised rays.

44. We have seen above (Art. 8*a*) that the simultaneous transmission of two plane-polarised rays of the same simple colour along the same line with the same velocity, but with different directions of planes of polarisation, has for general result an elliptically polarised ray of the same simple colour transmitted with the same velocity: further, the right-hand or left-hand character of the motion of the representative point round the ellipse depends only on the relation of the phases of the component rays. Hence, in general, an elliptically polarised ray, or, its special case, a circularly polarised ray, can be transmitted in any direction within a cubic crystal, or along the morphological axis of a tetragonal or hexagonal crystal; and its velocity is independent of its right-hand or left-hand character.

If the velocity of transmission of a plane-polarised ray along a given direction within a crystal is dependent on the azimuth of the plane of polarisation, we have seen (Art. 8*b*) that an elliptically or circularly polarised ray cannot result from the composition of two plane-polarised rays transmitted along that direction.

The transmission of a circularly polarised ray, however, may be possible even when that of a single plane-polarised ray is not so: for instance, a right-hand or a left-hand circularly polarised ray, but not a plane-polarised ray, can be transmitted along the morphological axis of a crystal of quartz. In such case, the velocities of transmission of a right-hand and a left-hand circularly polarised ray of the same simple colour are necessarily different: for it will be found on calculation that a right-hand and a left-hand circularly polarised ray *transmitted with the same velocity*, if superposed, are kinematically identical with a plane-polarised ray, the azimuth of the plane of polarisation of which depends solely on the relative phases of the component rays; but, according to hypothesis, a plane-polarised ray is incapable of transmission. Such a line will be an axis of optical symmetry, but cannot lie in a plane of general symmetry; for symmetry to the plane would require a right-hand and a left-hand ray to be transmissible with the same velocity.

In fact, if a right and left circular motion of the same radius and period are simultaneously impressed on the same particle, the resultant motion is a vibration along that diameter of the circle to which the two circular motions are symmetrical, namely, the diameter passing through the two positions of the particle which are identical for the component motions. If the two circular motions are transmitted through the medium with the same velocity, their relative phases, and thus the direction of the line of resultant vibration, will be the same at all points of the resultant ray: if they are transmitted with unequal velocities, the line of resultant vibration will have different azimuths for different points of the ray, and the change of azimuth will be proportional to the distance between the given points. Hence it follows that if a plane-polarised ray be incident normally on a plate cut perpendicularly to the morphological axis of a crystal of quartz, the ray will not be in a state of plane-polarisation within the plate, though it will be so after emergence: the planes of polarisation of the incident and emergent rays will be inclined to each other at an angle which is proportional to the thickness of the plate.

Summary.

1. Fresnel's hypothesis—that light consists in the vibratory motion of an incompressible elastic ether—being untenable, should be abandoned as an educational instrument.

2. The later hypothesis—that light consists in the vibratory motion of a compressible elastic ether, of which the elasticity (of volume and figure) is the same for all bodies and for all directions in the same body, and of which the effective density in bi-refractive media is dependent on the direction of the vibratory motion—satisfactorily accounts for most of the known optical laws: hence such terms as "axes of optical elasticity," which relate to variation of elasticity, must be discontinued.

3. Even this more satisfactory hypothesis may only be an approximate mechanical analogy, and may eventually be found to be inconsistent with experiment in some of its optical results; hence it cannot be satisfactorily used as the *basis* of a correlation of optical characters for the student of crystals; in fact, though it appears to be fully established that electro-magnetic waves and light-waves differ only in length, an electro-magnetic disturbance seems to be inexplicable as mere vibratory motion of an elastic body.

4. On the other hand, the accuracy of Huygens's construction is now so far confirmed by experiment that it doubtless expresses a Law of Nature.

5. This being the case, it is easily seen that the velocity and polarisation of each of the two rays transmissible in a given direction in a uniaxal crystal can be simply expressed by means of the *spheroid alone*:—

If R be a point on the spheroid, O the centre, RN the normal, NOr a line intersecting the normal perpendicularly, the point R corresponds to a ray transmissible in the direction NOr, with a velocity represented by constant $\frac{}{RN}$, and having its plane of polarisation perpendicular to RN.

6. Generalisation suggests that, in the case of crystals belonging to a lower type of general symmetry, there is a similar correspondence between each ray and a point on an *ellipsoid*.

7. Experiment confirms the rigorous accuracy of the generalisation.

8. The surface of reference, whether a sphere, spheroid or ellipsoid, may be conveniently denoted by the term *optical indicatrix*.

9. All the optical characters can be directly deduced from the indicatrix itself, and reference to its polar reciprocal is for this purpose unnecessary: further, it is possible to develop the characters from the consideration of *rays* alone.

10. The *front* of a pencil of rays which have started simultaneously from a point is part of the *ray-surface*; in the limit, if the pencil is of small aperture and includes a given ray, the pencil-front is part of the plane which touches the ray-surface where the ray meets it: hence, the pencil-front corresponding to the given ray may be briefly designated as the *ray-front*.

11. A plane passing through a ray and perpendicular to its plane of polarisation may be conveniently termed its *transverse* plane.

12. In such case, it follows that the normal to the ray-front corresponding to the ray Or lies in the transverse plane $RNOr$ and is perpendicular to OR, while the velocity of normal-propagation of the front is measured by $\frac{1}{OR}$.

13. The normal RN is the direction of vibration of the ray corresponding to the point R, if the most recent hypothesis as to the properties of an elastic luminiferous ether is true.

14. The so-called primary and secondary optic axes are not axes of symmetry, nor even constant lines, of the crystal: they may with precision be denoted respectively as the *optic bi-normals* and *bi-radials*; for they are directions in which the two normals drawn from the centre to tangent planes of the ray-surface having the same direction, or the two radii vectores of the ray-surface having the same direction, are respectively coincident with each other. A crystal may still be loosely termed *biaxal*, when it is merely desired to suggest that the interference-rings shown by a plate in convergent polarised light are rudely like those

which might be expected to be seen if the crystal had two axes, each identical in character with the optic axis of a tetragonal or hexagonal crystal.

15. By help of simple assumptions, which naturally present themselves and are consistent with all known experimental results, Fresnel's equation of the ray-surface may be deduced from the general principles of undulations, without regard to the physical character of the periodic change.

THE END.

Printed by Williams and Strahan, 7 Lawrence Lane, Cheapside.

www.ingramcontent.com/pod-product-compliance
Lightning Source LLC
Chambersburg PA
CBHW020122170426
43199CB00009B/606